Lóegaire Humphrey (Ed.)

Cumnoria

Lóegaire Humphrey (Ed.)

Cumnoria

Iguanodont, Dinosaur, Genus, Geologic time scale

Claud Press

Cover image: www.ingimage.com
Concerning the licence of the cover image please contact ingimage.

Publisher:
Claud Press is a trademark of
International Book Market Service Ltd., 17 Rue Meldrum, Beau Bassin, 1713-01 Mauritius
Email: info@bookmarketservice.com
Website: www.bookmarketservice.com

Published in 2011

Printed in: U.S.A., U.K., Germany. This book was not produced in Mauritius.

ISBN: 978-613-7-48684-9

Contents

Articles

Iguanodont	1
Cumnoria	4
Dinosaur	5
Genus	32
Geologic time scale	35
Kimmeridge Clay	42
George Rolleston	46

References

Article Sources and Contributors	49
Image Sources, Licenses and Contributors	51

Iguanodont

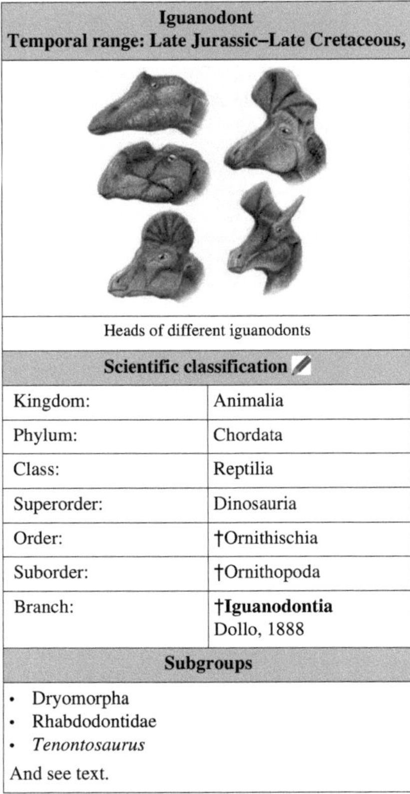

Iguanodont	
Temporal range: Late Jurassic–Late Cretaceous,	
Heads of different iguanodonts	
Scientific classification	
Kingdom:	Animalia
Phylum:	Chordata
Class:	Reptilia
Superorder:	Dinosauria
Order:	†Ornithischia
Suborder:	†Ornithopoda
Branch:	**†Iguanodontia** Dollo, 1888
Subgroups	
• Dryomorpha • Rhabdodontidae • *Tenontosaurus* And see text.	

Iguanodonts (members of the group **Iguanodontia**) were herbivorous dinosaurs that lived from the mid-Jurassic to Late Cretaceous. Some members include *Camptosaurus, Callovosaurus, Iguanodon, Ouranosaurus*, and the hadrosaurids or "duck-billed dinosaurs". Iguanodonts were one of the first groups of dinosaurs to be found. They are among the best known of the dinosaurs, and were among the most diverse and widespread herbivorous dinosaur groups of the Cretaceous period.[1] Iguanodontians were generally large animals, and some (such as *Shantungosaurus*, which measured up to 50 ft (15 m) in length and weighed up to 8 tons) equaled the largest carnivorous dinosaurs in size.

Classification

Iguanodontia is often listed as an infraorder within a suborder Ornithopoda, though Benton (2004) lists Ornithopoda as an infraorder and does not rank Iguanodontia. Traditionally, iguanodonts were grouped into the superfamily Iguanodontoidea and family Iguanodontidae. However, phylogenetic studies show that the traditional "iguanodontids" are a paraphyletic grade leading up to the hadrosaurs (duck-billed dinosaurs). Groups like Iguanodontoidea are sometimes still used as unranked clades in the scientific literature, though many traditional "iguanodontids" are now included in the more inclusive group Hadrosauroidea.

Iguanodontia is usually defined as the most inclusive group containing *Parasaurolophus walkeri* but not *Hypsilophodon foxii* or *Thescelosaurus neglectus*, or other combinations of species that would ultimately result in the same group in most modern analyses. The group was first defined as a clade in 2008 by Paul Sereno.[2]

Many iguanodonts have not yet been included in a large phylogenetic analyses, or are too fragmentary to place confidently. These include *Barilium*, *Bolong*, *Bihariosaurus*, *Delapparentia*, *Dollodon*, *Draconyx*, *Kukufeldia*,[3] *Hypselospinus*, *Macrogryphosaurus*, *Owenodon*, *Proplanicoxa*,[4] *Sellacoxa*[4] and *Xuwulong*. The simplified cladogram below follows an analysis by Andrew McDonald and colleagues, published in November 2010 with information from McDonald, 2011.[1] [5]

Rhabdodontidae

Tenontosaurus

Dryosauridae

Camptosaurus

Cumnoria

Uteodon

Hippodraco

Theiophytalia

Iguanodontia

Dryomorpha

Ankylopollexia

Styracosterna

Cedrorestes

Dakotadon

Iguanacolossus

Lanzhousaurus

Iguanodon

Hadrosauriformes

Mantellisaurus

Ouranosaurus

Hadrosauroidea

Cladogram after Butler *et al*, 2011.[6]

Talenkauen

Anabisetia

Rhabdodontidae

Iguanodontia   *Tenontosaurus*   *T. tilletti*

T. dossi

Dryosauridae

Ankylopollexia

References

[1] McDonald, A.T., Kirkland, J.I., DeBlieux, D.D., Madsen, S.K., Cavin, J., Milner, A.R.C. and Panzarin, L. (2010). "New Basal Iguanodonts from the Cedar Mountain Formation of Utah and the Evolution of Thumb-Spiked Dinosaurs." *PLoS ONE 5*, **11**: e14075. doi:10.1371/journal.pone.0014075

[2] Sereno, P.C. (2005). "Stem Archosauria Version 1.0." *TaxonSearch*. Available: (http://www.taxonsearch.org/Archive/stem-archosauria-1.0.php) via the Internet. Accessed 24 November 2010.

[3] McDonald, A.T., Barrett, P.M. and Chapman, S.D. (2010). "A new basal iguanodont (Dinosauria: Ornithischia) from the Wealden (Lower Cretaceous) of England." *Zootaxa*, **2569**: 1–43.

[4] Carpenter, K. and Ishida, Y. (2010). "Early and "Middle" Cretaceous Iguanodonts in Time and Space" (http://www.ucm.es/info/estratig/ JIG/vol_content/vol_36_2/36_2_145_164_Carpenter.pdf). *Journal of Iberian Geology* **36** (2): 145–164. doi:10.5209/rev_JIGE.2010.v36.n2.3. .

[5] Andrew T. McDonald (2011). "The taxonomy of species assigned to *Camptosaurus* (Dinosauria: Ornithopoda)" (http://www.mapress.com/ zootaxa/2011/f/z02783p068f.pdf). *Zootaxa* **2783**: 52–68. .

[6] Richard J. Butler, Jin Liyong, Chen Jun, Pascal Godefroit (2011). "The postcranial osteology and phylogenetic position of the small ornithischian dinosaur *Changchunsaurus parvus* from the Quantou Formation (Cretaceous: Aptian–Cenomanian) of Jilin Province, north-eastern China". *Palaeontology* **54** (3): 667–683. doi:10.1111/j.1475-4983.2011.01046.x.

Cumnoria

Cumnoria is a genus of herbivorous iguanodontian dinosaur. It is a basal iguanodontian which lived during the upper Jurassic period (Kimmeridgian age) in what is now Oxfordshire, United Kingdom.

Cumnoria is known from the holotype OXFUM J.3303, a partial skull and postcranium, recovered from the lower Kimmeridge Clay Formation, in the Chawley Brick Pits, Cumnor Hurst. Workers at first discarded the remains on a dump heap, but one of them later collected the bones in a sack and showed them to Professor George Rolleston, an anatomist at the nearby Oxford University. Rolleston in turn brought them to the attention of palaeontologist Professor Joseph Prestwich who in 1879 reported them as a new species of *Iguanodon*, though without actually coining a species name.[1] In 1880 Prestwich published an article on the geological stratigraphy of the find.[2] The same year John Whitaker Hulke named the species *Iguanodon prestwichii*, the specific epithet honouring Prestwich.[3]

In 1888, Harry Govier Seeley decided the taxon represented a separate and new genus which he named *Cumnoria* after Cumnor. Its type species *Iguanodon prestwichii* was thus recombined into *Cumnoria prestwichii* — though Seeley spelled the epithet as *prestwichi*.[4] The genus was quickly abandoned however: already in 1889 Richard Lydekker assigned the species to *Camptosaurus*, as *Camptosaurus prestwichii*.[5] This opinion was generally accepted for over a century. In 1980 Peter Galton provided the first modern description of the species.[6]

In 1998 David Norman concluded that Seeley's original generic distinction was valid.[7] In 2008 this was supported by Darren Naish and David Martill.[8] In 2010 and 2011 cladistic analyses by Andrew T. McDonald confirmed this by showing that *Cumnoria* had a separate phylogenetic position from *Camptosaurus dispar*.[9] [10]

The holotype of *Cumnoria* is of a rather small bipedal animal, with a gracile build, about 3.5 metres long. The specimen is probably that of a juvenile though.[6]

Camptosaurus prestwichii was traditionally assigned to the Camptosauridae. In the new analyses of McDonald *Cumnoria* has instead been recovered as a basal member of the Styracosterna, more closely related to more derived ("advanced") iguanodontians than to *Camptosaurus dispar*. *Cumnoria* would then be the oldest known styracostern.[10]

References

[1] J. Prestwich. 1879. "On the discovery of a species of *Iguanodon* in the Kimmeridge Clay near Oxford; and a notice of a very fossiliferous band of the Shotover Sands", *Geological Magazine, new series, decade 2* **6**(5): 193-195

[2] Prestwich, J., 1880, "Note on the occurrence of a new species of *Iguanodon* in a brickpit of the Kimmeridge Clay at Cumnor Hurst, three miles W.S.W. of Oxford", *Quarterly Journal of the Geological Society of London*, **36**: 430-432

[3] J.W. Hulke, 1880, "*Iguanodon prestwichii*, a new species from the Kimmeridge Clay, distinguished from *I. mantelli* of the Wealden Formation in the S.E. of England and Isle of Wight by differences in the shape of the vertebral centra, by fewer than five sacral vertebrae, by the simpler character of its tooth-serrature, etc., founded on numerous fossil remains lately discovered at Cumnor, near Oxford", *Quarterly Journal of the Geological Society of London* **36**(143):433-456

[4] H.G. Seeley, 1888, "On *Cumnoria*, an iguanodont genus founded upon the *Iguanodon prestwichi*, Hulke", *Report of the British Association for the Advancement of Science* **57**:698

[5] Lydekker, R., 1889, "On the remains and affinities of five genera of Mesozoic reptiles", *Quarterly Journal of the Geological Society*, **45**: 41-59

[6] Galton, P.M. and Powell, H.P., 1980, "The ornithischian dinosaur *Camptosaurus prestwichii* from the Upper Jurassic of England", *Palaeontology*, **23**: 411-443

[7] Norman, D., 1998, "On Asian ornithopods (Dinosauria: Ornithischia). 3. A new species of iguanodontid dinosaur", *Zoological Journal of the Linnean Society* **122**: 291-348

[8] Naish, D. and Martill, D.M., 2008, "Dinosaurs of Great Britain and the rôle of the Geological Society of London in their discovery: Ornithischia", *Journal of the Geological Society of London*, **165**: 613-623

[9] McDonald, A.T., Kirkland, J.I., DeBlieux, D.D., Madsen, S.K., Cavin, J., Milner, A.R.C. and Panzarin, L. (2010). "New Basal Iguanodonts from the Cedar Mountain Formation of Utah and the Evolution of Thumb-Spiked Dinosaurs". *PLoS ONE* **5** (11): e14075. doi:10.1371/journal.pone.0014075.

[10] Andrew T. McDonald (2011). "The taxonomy of species assigned to *Camptosaurus* (Dinosauria: Ornithopoda)" (http://www.mapress. com/zootaxa/2011/f/z02783p068f.pdf). *Zootaxa* **2783**: 52–68. .

Dinosaur

Dinosaurs (from Greek: δεινός terrible or potent, and σαύρα lizard) are a diverse group of animals that were the dominant terrestrial vertebrates for over 160 million years, from the late Triassic period (about 230 million years ago) until the end of the Cretaceous (about 65 million years ago), when the Cretaceous–Paleogene extinction event led to the extinction of most dinosaur species at the close of the Mesozoic era. The fossil record indicates that birds evolved within theropod dinosaurs during the Jurassic period. Some of them survived the Cretaceous–Paleogene extinction event, including the ancestors of all modern birds. Consequently, in modern classification systems, birds are considered a type of dinosaur—the only group which survived to the present day.[1] [2]

Dinosaurs are a diverse and varied group of animals; birds, at over 9,000 species, are the most diverse group of vertebrate besides perciform fish.[3] Paleontologists have identified over 500 distinct genera[4] and more than 1,000 different species of non-avian dinosaurs.[5] Dinosaurs are represented on every continent by both extant species and fossil remains.[6] Some dinosaurs are herbivorous, others carnivorous. Many dinosaurs, including birds, have been bipedal, though many extinct groups were quadrupedal, and some were able to shift between these body postures. Many species possess elaborate display structures such as horns or crests, and some prehistoric groups developed even more elaborate skeletal modifications such as bony armor. Avian dinosaurs have been the planet's dominant flying vertebrate since the extinction of the pterosaurs, and evidence suggests that all ancient dinosaurs built nests and laid eggs much as avian species do today. Although generally known for the large size of some species, most Mesozoic dinosaurs were human-sized or even smaller.

The term "dinosaur" was coined in 1842 by the English paleontologist Richard Owen, and derives from Greek δεινός (*deinos*) "terrible, powerful, wondrous, potent" + σαῦρος (*sauros*) "lizard". Through the first half of the 20th century, most of the scientific community believed dinosaurs to have been sluggish, unintelligent cold-blooded animals. Most research conducted since the 1970s, however, has indicated that dinosaurs were active animals with elevated metabolisms and numerous adaptations for social interaction.

Since the first dinosaur fossils were recognized in the early 19th century, mounted fossil dinosaur skeletons or replicas have been major attractions at museums around the world, and dinosaurs have become a part of world culture. They have been featured in best-selling books and films such as *Jurassic Park*, and new discoveries are regularly covered by the media. In informal speech, the word "dinosaur" is used to describe things that are impractically large, obsolete, or bound for extinction,[7] reflecting the outdated view that dinosaurs were maladapted monsters of the ancient world.

Name

The taxon **Dinosauria** was formally named in 1842 by Sir Richard Owen, who used it to refer to the "distinct tribe or sub-order of Saurian Reptiles" that were then being recognized in England and around the world.[8] :103 The term is derived from the Greek words δεινός (*deinos* meaning "terrible", "powerful", or "wondrous") and σαῦρος (*sauros* meaning "lizard" or "reptile").[8] :103[9] Though the taxonomic name has often been interpreted as a reference to dinosaurs' teeth, claws, and other fearsome characteristics, Owen intended it merely to evoke their size and majesty.[10]

Modern definition

Formal definitions are written to correspond with scientific conceptions of dinosaurs that predate the modern use of phylogenetics. The continuity of meaning is intended to prevent confusion about what the term "dinosaur" means.

Under phylogenetic taxonomy, dinosaurs are usually defined as the group consisting of "*Triceratops*, Neornithes [modern birds], their most recent common ancestor, and all descendants".[11] It has also been suggested that Dinosauria be defined with respect to the MRCA of *Megalosaurus* and *Iguanodon*, because these were two of the three genera cited by Richard Owen when he recognized the Dinosauria.[12] Both definitions result in the same set of animals being defined as dinosaurs, that is "Dinosauria = Ornithischia + Saurischia", which

Triceratops skeleton at the American Museum of Natural History in New York City

encompasses theropods (mostly bipedal carnivores and birds), ankylosaurians (armored herbivorous quadrupeds), stegosaurians (plated herbivorous quadrupeds), ceratopsians (herbivorous quadrupeds with horns and frills), ornithopods (bipedal or quadrupedal herbivores including "duck-bills"), and presumptuously, sauropodomorphs (mostly large herbivorous quadrupeds with long necks and tails).

Many paleontologists note that the order in which sauropodomorphs and theropods diverged may omit sauropodomorphs from the definition for both saurischians and dinosaurs. To avoid the instability of Dinosauria, a more conservative definition of Dinosauria is defined with respect to four anchoring nodes: *Triceratops horridus*, *Saltasaurus loricatus*, and *Passer domesticus*, their most recent common ancestor, and all descendants. This "safer" definition can be expressed as "Dinosauria = Ornithischia + Sauropodomorpha + Theropoda".[13]

There is a wide consensus among paleontologists that birds are the descendants of theropod dinosaurs. Using the strict phylogenetic nomenclatural definition that all descendants of a single common ancestor must be included in a group for that group to be natural, birds would thus *be* dinosaurs and dinosaurs are, therefore, not extinct. Birds are classified by most paleontologists as belonging to the subgroup Maniraptora, which are coelurosaurs, which are theropods, which are saurischians, which are dinosaurs.[14]

From the point of view of cladistics, birds are dinosaurs, but in ordinary speech the word "dinosaur" does not include birds. Additionally, referring to dinosaurs that are not birds as "non-avian dinosaurs" is cumbersome. For clarity, this article will use "dinosaur" as a synonym for "non-avian dinosaur". The term "non-avian dinosaur" will be used for emphasis as needed.

General description

Using one of the above definitions, dinosaurs (aside from birds) can be generally described as terrestrial archosaurian reptiles with limbs held erect beneath the body, that existed from the Late Triassic (first appearing in the Carnian faunal stage) to the Late Cretaceous (going extinct at the end of the Maastrichtian).[15] Many prehistoric animals are popularly conceived of as dinosaurs, such as ichthyosaurs, mosasaurs, plesiosaurs, pterosaurs, and *Dimetrodon*, but are not classified scientifically as dinosaurs. Marine reptiles like ichthyosaurs, mosasaurs, and plesiosaurs were neither terrestrial nor archosaurs; pterosaurs were archosaurs but not terrestrial; and *Dimetrodon* was a

Stegosaurus skeleton, Field Museum, Chicago

Permian animal more closely related to mammals.[16] Dinosaurs were the dominant terrestrial vertebrates of the Mesozoic, especially the Jurassic and Cretaceous. Other groups of animals were restricted in size and niches; mammals, for example, rarely exceeded the size of a cat, and were generally rodent-sized carnivores of small prey.[17] One notable exception is *Repenomamus giganticus*, a triconodont weighing between 12 kilograms (26 lb) and 14 kilograms (31 lb) that is known to have eaten small dinosaurs like young *Psittacosaurus*.[18]

Dinosaurs were an extremely varied group of animals; according to a 2006 study, over 500 dinosaur genera have been identified with certainty so far, and the total number of genera preserved in the fossil record has been estimated at around 1850, nearly 75% of which remain to be discovered.[4] An earlier study predicted that about 3400 dinosaur genera existed, including many which would not have been preserved in the fossil record.[19] As of September 17, 2008, 1047 different species of dinosaurs have been named.[5] Some were herbivorous, others carnivorous. Some dinosaurs were bipeds, some were quadrupeds, and others, such as *Ammosaurus* and *Iguanodon*, could walk just as easily on two or four legs. Many had bony armor, or cranial modifications like horns and crests. Although known for large size, many dinosaurs were human-sized or smaller. Dinosaur remains have been found on every continent on Earth, including Antarctica.[6] No non-avian dinosaurs are known to have lived in marine habitats or in aerial habitats, although it is possible some feathered non-avian theropods were flyers. There is also evidence that some spinosaurids had semi-aquatic habits.[20]

Distinguishing anatomical features

While recent discoveries have made it more difficult to present a universally agreed-upon list of dinosaurs' distinguishing features, nearly all dinosaurs discovered so far share certain modifications to the ancestral archosaurian skeleton. Although some later groups of dinosaurs featured further modified versions of these traits, they are considered typical across Dinosauria; the earliest dinosaurs had them and passed them on to all their descendants. Such common features across a taxonomic group are called synapomorphies.

A detailed assessment of archosaur interrelations by S. Nesbitt[21] confirmed or found the following 12 unambiguous synapomorphies, some previously known:

- in the skull, a supratemporal fossa (excavation) is present in front of the supratemporal fenestra
- epipophyses present in anterior neck vertebrae (except atlas and axis)
- apex of deltopectoral crest (a projection on which the deltopectoral muscles attach) located at or more than 30% down the length of the humerus (upper arm bone)
- radius shorter than 80% of humerus length
- fourth trochanter (projection where the caudofemoralis muscle attaches) on the femur (thigh bone) is a sharp flange
- fourth trochanter asymmetrical, with distal margin forming a steeper angle to the shaft
- on the astragalus and calcaneum the proximal articular facet for fibula occupies less than 30% of the transverse width of the element
- exocciptials (bones at the back of the skull) do not meet along the midline on the floor of the endocranial cavity
- proximal articular surfaces of the ischium with the ilium and the pubis separated by a large concave surface
- cnemial crest on the tibia (shinbone) arcs anterolaterally
- distinct proximodistally oriented ridge present on the posterior face of the distal end of the tibia

Nesbitt found a number of further potential synapomorphies, and discounted a number of synapomorphies previously suggested. Some of these are also present in silesaurids, which Nesbitt recovered as a sister group to Dinosauria, including a large anterior trochanter, metatarsals II and IV of subequal length, reduced contact between ischium and pubis, the presence of a cenmial crest on the tibia and of an ascending process on the astragalus,[11] and many others.

A variety of other skeletal features were shared by many dinosaurs. However, because they were either common to other groups of archosaurs or were not present in all early dinosaurs, these features are not considered to be synapomorphies. For example, as diapsid reptiles, dinosaurs ancestrally had two pairs of temporal fenestrae (openings in the skull behind the eyes), and as members of the diapsid group Archosauria, had additional openings in the snout and lower jaw.[22] Additionally, several characteristics once thought to be synapomorphies are now known to have appeared before dinosaurs, or were absent in the earliest dinosaurs and independently evolved by

Edmontonia was an armored dinosaur of the group Ankylosauria.

different dinosaur groups. These include an elongated scapula, or shoulder blade; a sacrum composed of three or more fused vertebrae (three are found in some other archosaurs, but only two are found in *Herrerasaurus*);[11] and an acetabulum, or hip socket, with a hole at the center of its inside surface (closed in *Saturnalia*, for example).[23] Another difficulty of determining distinctly dinosaurian features is that early dinosaurs and other archosaurs from the Late Triassic are often poorly known and were similar in many ways; these animals have sometimes been misidentified in the literature.[24]

Dinosaurs stood erect in a manner similar to most modern mammals, but distinct from most other reptiles, whose limbs sprawl out to either side.[25] Their posture was due to the development of a laterally facing recess in the pelvis (usually an open socket) and a corresponding inwardly facing distinct head on the femur.[26] Their erect posture enabled dinosaurs to breathe easily while moving, which likely permitted stamina and activity levels that surpassed those of "sprawling"

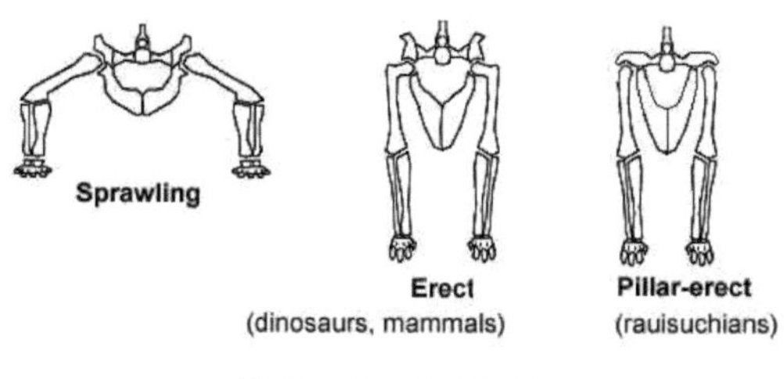

Hip joints and hindlimb postures

reptiles.[27] Erect limbs probably also helped support the evolution of large size by reducing bending stresses on limbs.[28] Some non-dinosaurian archosaurs, including rauisuchians, also had erect limbs but achieved this by a "pillar erect" configuration of the hip joint, where instead of having a projection from the femur insert on a socket on the hip, the upper pelvic bone was rotated to form an overhanging shelf.[28]

Natural history

Origins and early evolution

For a long time many scientists thought dinosaurs were polyphyletic with multiple groups of unrelated "dinosaurs" evolving due to similar pressures,[29] [30] [31] but dinosaurs are now known to have formed a single group.[11] [21] [32] Dinosaurs diverged from their archosaur ancestors approximately 230 million years ago during the Middle to Late Triassic period, roughly 20 million years after the Permian–Triassic extinction event wiped out an estimated 95% of all life on Earth.[33] [34] Radiometric dating of the rock formation that

Marasuchus, a dinosaur-like ornithodiran

contained fossils from the early dinosaur genus *Eoraptor* establishes its presence in the fossil record at this time. Paleontologists believe *Eoraptor* resembles the common ancestor of all dinosaurs;[35] if this is true, its traits suggest that the first dinosaurs were small, bipedal predators.[36] The discovery of primitive, dinosaur-like ornithodirans such as *Marasuchus* and *Lagerpeton* in Argentinian Middle Triassic strata supports this view; analysis of recovered fossils suggests that these animals were indeed small, bipedal predators.

When dinosaurs appeared, terrestrial habitats were occupied by various types of basal archosaurs and therapsids, such as aetosaurs, cynodonts, dicynodonts, ornithosuchids, rauisuchias, and rhynchosaurs. Most of these other animals became extinct in the Triassic, in one of two events. First, at about the boundary between the Carnian and Norian faunal stages (about 215 million years ago), dicynodonts and a variety of basal archosauromorphs, including the prolacertiforms and rhynchosaurs, became extinct. This was followed by the Triassic–Jurassic extinction event (about 200 million years ago), that saw the end of most of the other groups of early archosaurs, like aetosaurs, ornithosuchids, phytosaurs, and rauisuchians. These losses left behind a land fauna of crocodylomorphs, dinosaurs, mammals, pterosaurians, and turtles.[11]

The first few lines of primitive dinosaurs diversified through the Carnian and Norian stages of the Triassic, most likely by occupying the niches of groups that became extinct. Traditionally, dinosaurs were thought to have replaced the variety of other Triassic land animals by proving superior through a long period of competition. This now appears unlikely, for several reasons. Dinosaurs do not show a pattern of steadily increasing in diversity and numbers, as would be predicted if they were competitively replacing other groups; instead, they were very rare through the Carnian, making up only 1–2% of individuals present in faunas. In the Norian, however, after the extinction of

The early forms *Herrerasaurus* (large), *Eoraptor* (small) and a *Plateosaurus* skull

several other groups, they became significant components of faunas, representing 50–90% of individuals. Also, what had been viewed as a key adaptation of dinosaurs, their erect stance, is now known to have been present in several contemporaneous groups that were not as successful (aetosaurs, ornithosuchids, rauisuchians, and some groups of crocodylomorphs). Finally, the Late Triassic itself was a time of great upheaval in life, with shifts in plant life, marine life, and climate.[11] Crurotarsans, today represented only by crocodilians but in the Late Triassic also encompassing such now-extinct groups as aetosaurs, phytosaurs, ornithosuchians, and rauisuchians, were actually more diverse in the Late Triassic than dinosaurs, indicating that the survival of dinosaurs had more to do with luck than superiority.[37]

Classification

Dinosaurs (including birds) are archosaurs, like modern crocodilians. Archosaurs' diapsid skulls have two holes, called temporal fenestrae, located where the jaw muscles attach, and an additional antorbital fenestra in front of the eyes. Most reptiles (including birds) are diapsids; mammals, with only one temporal fenestra, are called synapsids; and turtles, with no temporal fenestra, are anapsids. Anatomically, dinosaurs share many other archosaur characteristics, including teeth that grow from sockets rather than as direct extensions of the jawbones. Within the archosaur group, dinosaurs are differentiated most noticeably by their gait. Dinosaur legs extend directly beneath the body, whereas the legs of lizards and crocodilians sprawl out to either side.

Collectively, dinosaurs are usually regarded as a superorder or an unranked clade. They are divided into two orders, Saurischia and Ornithischia, depending upon pelvic structure. Saurischia includes those taxa sharing a more recent common ancestor with birds than with Ornithischia, while Ornithischia includes all taxa sharing a more recent common ancestor with *Triceratops* than with Saurischia. Saurischians ("lizard-hipped", from the Greek *sauros* (σαυρος) meaning "lizard" and *ischion* (ισχιον) meaning "hip joint") retained the hip structure of their ancestors, with a pubis bone directed cranially, or forward.[26] This basic form was modified by rotating the pubis backward to varying degrees in several groups (*Herrerasaurus*,[38] therizinosauroids,[39] dromaeosaurids,[40] and birds[14]). Saurischia includes the theropods (bipedal and mostly carnivores, except for birds) and sauropodomorphs (long-necked quadrupedal herbivores).

By contrast, ornithischians ("bird-hipped", from the Greek *ornitheios* (ορνιθειος) meaning "of a bird" and *ischion* (ισχιον) meaning "hip joint") had a pelvis that superficially resembled a bird's pelvis: the pubis bone was oriented caudally (rear-pointing) Unlike birds, the ornithischian pubis also usually had an additional forward-pointing process. Ornithischia includes a variety of herbivores. (**NB:** the terms "lizard hip" and "bird hip" are misnomers – birds evolved from dinosaurs with "lizard hips".)

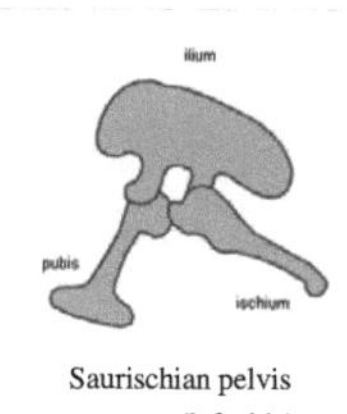

Saurischian pelvis structure (left side)

Tyrannosaurus pelvis (showing saurischian structure – left side)

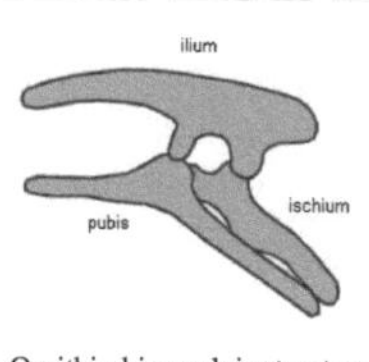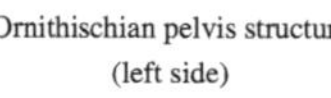

Ornithischian pelvis structure (left side)

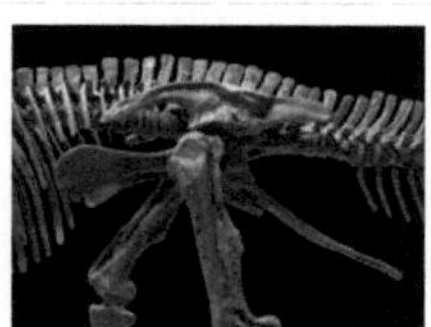

Edmontosaurus pelvis (showing ornithischian structure – left side)

The following is a simplified classification of dinosaur families. A more detailed version can be found at List of dinosaur classifications.

- Dinosauria
    - Saurischia (theropods and sauropods)
        - †Herrerasaurians (early bipedal predators)
        - Theropods (all bipedal; most were carnivores)
            - †Coelophysoids (*Coelophysis* and close relatives)
            - †Ceratosaurians (*Ceratosaurus* and abelisaurids – the latter were important Late Cretaceous predators in southern continents)
            - †Spinosauroids (long bodies; short arms; some with crocodile-like skulls and bony "sails" on their backs)
            - †Carnosaurians (*Allosaurus* and close relatives, like *Carcharodontosaurus*)
            - Coelurosaurians (diverse, with a range of body sizes and niches)
                - †Tyrannosauroids (small to gigantic, often with reduced forelimbs)
                - †Ornithomimosaurians ("ostrich-mimics"; mostly toothless; carnivores to possible herbivores)
                - †Therizinosauroids (bipedal herbivores with large hand claws and small heads)
                - †Oviraptorosaurians (mostly toothless; their diet and lifestyle are uncertain)
                - †Dromaeosaurids (popularly known as "raptors"; bird-like carnivores)
                - †Troodontids (similar to dromaeosaurids, but more lightly built)
                - Avialans (flying dinosaurs, including modern birds: the only living dinosaurs)

Several macronarian Sauropods: from left to right *Camarasaurus*, *Brachiosaurus*, *Giraffatitan*, and *Euhelopus*

Various ornithopod dinosaurs and one heterodontosaurid. Far left: *Camptosaurus*, left: *Iguanodon*, center background: *Shantungosaurus*, center foreground: *Dryosaurus*, right: *Corythosaurus*, far right (small): *Heterodontosaurus*, far right (large) *Tenontosaurus*.

- †Sauropodomorphs (quadrupedal herbivores with small heads, long necks and tails, and elephant-like bodies)
  - †"Prosauropods" (early relatives of sauropods; small to quite large; some possibly omnivorous; bipeds and quadrupeds)
  - †Sauropods (very large, usually over 15 meters long [49 ft])
    - †Diplodocoids (skulls and tails elongated; teeth typically narrow and pencil-like)
    - †Macronarians (boxy skulls; spoon- or pencil-shaped teeth)
      - †Brachiosaurids (very long necks; forelimbs longer than hindlimbs)
      - †Titanosaurians (diverse; stocky, with wide hips; most common in the Late Cretaceous of southern continents)
- †Ornithischians (diverse bipedal and quadrupedal herbivores)
  - †Heterodontosaurids (meter- or yard-scale herbivores or omnivores with prominent canine teeth)
  - †Thyreophorans (armored dinosaurs; mostly quadrupeds)
    - †Ankylosaurians (scutes as primary armor; some had club-like tails)
    - †Stegosaurians (spikes and plates as primary armor)
  - †Ornithopods (diverse, from meter- or yard-scale bipeds to 12-meter (39 ft) animals that could move as both bipeds and quadrupeds; evolved a method of chewing using skull flexibility and large numbers of teeth)
    - †Hadrosaurids ("duckbilled dinosaurs")
  - †Pachycephalosaurians ("bone-heads"; bipeds with domed or knobby growth on skulls)
  - †Ceratopsians (dinosaurs with horns and frills, although most early forms had only the beginnings of these features)

Evolution and paleobiogeography

Dinosaur evolution after the Triassic follows changes in vegetation and the location of continents. In the Late Triassic and Early Jurassic, the continents were connected as the single landmass Pangaea, and there was a worldwide dinosaur fauna mostly composed of coelophysoid carnivores and prosauropod herbivores.[41] Gymnosperm plants (particularly conifers), a potential food source, radiated in the Late Triassic. Prosauropods did not have sophisticated mechanisms for processing food in the mouth, and so must have employed other means of breaking down food farther along the digestive tract.[42] The general homogeneity of dinosaurian faunas continued into the Middle and Late Jurassic, where most localities had predators consisting of ceratosaurians, spinosauroids, and carnosaurians, and herbivores consisting of stegosaurian ornithischians and large sauropods. Examples of this include the Morrison Formation of North America and Tendaguru Beds of Tanzania. Dinosaurs in China show some differences, with specialized sinraptorid theropods and unusual, long-necked sauropods like *Mamenchisaurus*.[41] Ankylosaurians and ornithopods were also becoming more common, but prosauropods had become extinct. Conifers and pteridophytes were the most common plants. Sauropods, like the earlier prosauropods, were not oral processors, but ornithischians were evolving various means of dealing with food in the mouth, including potential cheek-like organs to keep food in the mouth, and jaw motions to grind food.[42] Another notable evolutionary event of the Jurassic was the appearance of true birds, descended from maniraptoran coelurosaurians.[14]

By the Early Cretaceous and the ongoing breakup of Pangaea, dinosaurs were becoming strongly differentiated by landmass. The earliest part of this time saw the spread of ankylosaurians, iguanodontians, and brachiosaurids through Europe, North America, and northern Africa. These were later supplemented or replaced in Africa by large spinosaurid and carcharodontosaurid theropods, and rebbachisaurid and titanosaurian sauropods, also found in South America. In Asia, maniraptoran coelurosaurians like dromaeosaurids, troodontids, and oviraptorosaurians became the common theropods, and ankylosaurids and early ceratopsians like *Psittacosaurus* became important herbivores.

Meanwhile, Australia was home to a fauna of basal ankylosaurians, hypsilophodonts, and iguanodontians.[41] The stegosaurians appear to have gone extinct at some point in the late Early Cretaceous or early Late Cretaceous. A major change in the Early Cretaceous, which would be amplified in the Late Cretaceous, was the evolution of flowering plants. At the same time, several groups of dinosaurian herbivores evolved more sophisticated ways to orally process food. Ceratopsians developed a method of slicing with teeth stacked on each other in batteries, and iguanodontians refined a method of grinding with tooth batteries, taken to its extreme in hadrosaurids.[42] Some sauropods also evolved tooth batteries, best exemplified by the rebbachisaurid *Nigersaurus*.[43]

There were three general dinosaur faunas in the Late Cretaceous. In the northern continents of North America and Asia, the major theropods were tyrannosaurids and various types of smaller maniraptoran theropods, with a predominantly ornithischian herbivore assemblage of hadrosaurids, ceratopsians, ankylosaurids, and pachycephalosaurians. In the southern continents that had made up the now-splitting Gondwana, abelisaurids were the common theropods, and titanosaurian sauropods the common herbivores. Finally, in Europe, dromaeosaurids, rhabdodontid iguanodontians, nodosaurid ankylosaurians, and titanosaurian sauropods were prevalent.[41] Flowering plants were greatly radiating,[42] with the first grasses appearing by the end of the Cretaceous.[44] Grinding hadrosaurids and shearing ceratopsians became extremely diverse across North America and Asia. Theropods were also radiating as herbivores or omnivores, with therizinosaurians and ornithomimosaurians becoming common.[42]

The Cretaceous–Paleogene extinction event, which occurred approximately 65 million years ago at the end of the Cretaceous period, caused the extinction of all dinosaurs except for the birds. Some other diapsid groups, such as crocodilians, lizards, snakes, sphenodontians, and choristoderans, also survived the event.[45]

Paleobiology

Knowledge about dinosaurs is derived from a variety of fossil and non-fossil records, including fossilized bones, feces, trackways, gastroliths, feathers, impressions of skin, internal organs and soft tissues.[46] [47] Many fields of study contribute to our understanding of dinosaurs, including physics (especially biomechanics), chemistry, biology, and the earth sciences (of which paleontology is a sub-discipline). Two topics of particular interest and study have been dinosaur size and behavior.

Size

The sauropods were the largest dinosaurs. For much of the dinosaur era, the smallest sauropods were larger than anything else in their habitat, and the largest were an order of magnitude more massive than anything else that has since walked the Earth. Giant

Scale diagram comparing the largest known dinosaurs in five major clades and a human

prehistoric mammals such as the *Paraceratherium* and the Columbian mammoth were dwarfed by the giant sauropods, and only a handful of modern aquatic animals approach or surpass them in size – most notably the blue whale, which reaches up to 173000 kg (381000 lb) and over 30 meters (98 ft) in length. There are several proposed advantages for the large size of sauropods, including protection from predation, reduction of energy use, and longevity, but it may be that the most important advantage was dietary. Large animals are more efficient at digestion than small animals, because food spends more time in their digestive systems. This also permits them to subsist on food with lower nutritive value than smaller animals. Sauropod remains are mostly found in rock formations interpreted as dry or seasonally dry, and the ability to eat large quantities of low-nutrient browse would have been advantageous in such environments.[48]

Most dinosaurs, however, were much smaller than the giant sauropods. Current evidence suggests that dinosaur average size varied through the Triassic, early Jurassic, late Jurassic and Cretaceous periods.[35] Theropod dinosaurs, when sorted by estimated weight into categories based on order of magnitude, most often fall into the 100 to 1000 kilogram (220 to 2200 lb) category, whereas recent predatory carnivorans peak in the 10 to 100 kilogram (22 to 220 lb) category.[49] The mode of dinosaur body masses is between one and ten metric tonnes.[50] This contrasts sharply with the size of Cenozoic mammals, estimated by the National Museum of Natural History as about 2 to 5 kilograms (5 to 10 lb).[51]

Largest and smallest

Only a tiny percentage of animals ever fossilize, and most of these remain buried in the earth. Few of the specimens that are recovered are complete skeletons, and impressions of skin and other soft tissues are rare. Rebuilding a complete skeleton by comparing the size and morphology of bones to those of similar, better-known species is an inexact art, and reconstructing the muscles and other organs of the living animal is, at best, a process of educated guesswork. As a result, scientists will probably never be certain of the largest and smallest dinosaurs.

The tallest and heaviest dinosaur known from good skeletons is *Giraffatitan brancai* (previously classified as a species of *Brachiosaurus*). Its remains were discovered in Tanzania between 1907–12. Bones from several similar-sized individuals were incorporated into the skeleton now mounted and on display at the Museum für Naturkunde Berlin;[52] this mount is 12 meters (39 ft) tall and 22.5 meters (74 ft) long, and would have belonged to an animal that weighed between 30000 and 60000 kilograms (70000 and 130000 lb). The longest complete dinosaur is the 27-meter (89 ft) long

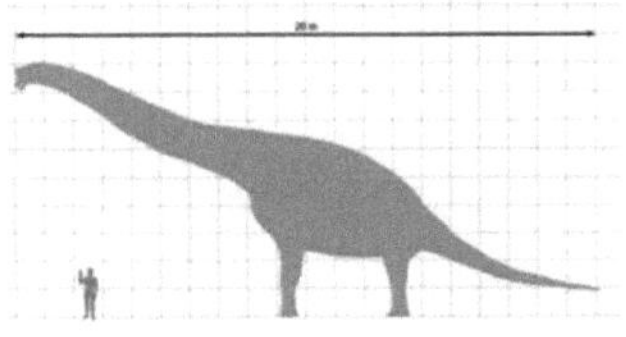
Comparative size of *Giraffatitan*

Diplodocus, which was discovered in Wyoming in the United States and displayed in Pittsburgh's Carnegie Natural History Museum in 1907.

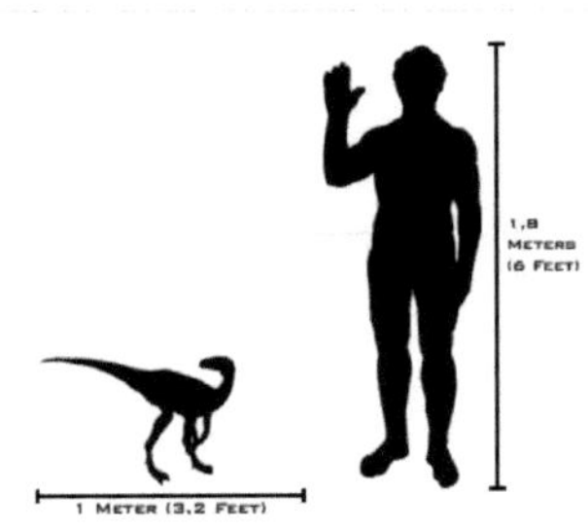

Comparative size of *Eoraptor*

There were larger dinosaurs, but knowledge of them is based entirely on a small number of fragmentary fossils. Most of the largest herbivorous specimens on record were all discovered in the 1970s or later, and include the massive *Argentinosaurus*, which may have weighed 80000 to 100000 kilograms (90 to 110 short tons); some of the longest were the 33.5 meters (110 ft) long *Diplodocus hallorum*[48] (formerly *Seismosaurus*) and the 33 meters (108 ft) long *Supersaurus*;[53] and the tallest, the 18 meters (59 ft) tall *Sauroposeidon*, which could have reached a sixth-floor window. The heaviest and longest of them all may have been *Amphicoelias fragillimus*, known only from a now lost partial vertebral neural arch described in 1878. Extrapolating from the illustration of this bone, the animal may have been 58 meters (190 ft) long and weighed over 120000 kg (260000 lb).[48] The largest known carnivorous dinosaur was *Spinosaurus*, reaching a length of 16 to 18 meters (50 to 60 ft), and weighing in at 8150 kg (18000 lb).[54] Other large meat-eaters included *Giganotosaurus*, *Carcharodontosaurus* and *Tyrannosaurus*.[55]

Not including modern birds, the smallest known dinosaurs known were about the size of a pigeon.[56] The theropods *Anchiornis* and *Epidexipteryx* both had a total skeletal length of under 35 centimeters (1.1 ft).[56] [57] *Anchiornis* is currently the smallest dinosaur described from an adult specimen, with an estimated weight of 110 grams.[57] The smallest herbivorous dinosaurs included *Microceratus* and *Wannanosaurus*, at about 60 cm (2 ft) long each.[58] [59]

Behavior

Interpretations of dinosaur behavior are generally based on the pose of body fossils and their habitat, computer simulations of their biomechanics, and comparisons with modern animals in similar ecological niches. As such, the current understanding of dinosaur behavior relies on speculation, and will likely remain controversial for the foreseeable future. However, there is general agreement that some behaviors which are common in crocodiles and birds, dinosaurs' closest living relatives, were also common among dinosaurs.

A nesting ground of *Maiasaura* was discovered in 1978.

The first potential evidence of herding behavior was the 1878 discovery of 31 *Iguanodon* dinosaurs which were then thought to have perished together in Bernissart, Belgium, after they fell into a deep, flooded sinkhole and drowned.[60] Other mass-death sites have been subsequently discovered. Those, along with multiple trackways, suggest that gregarious behavior was common in many dinosaur species. Trackways of hundreds or even thousands of herbivores indicate that duck-bills (hadrosaurids) may have moved in great herds, like the American Bison or the African Springbok. Sauropod tracks document that these animals traveled in groups composed of several different species, at least in Oxfordshire, England,[61] although there is not evidence for specific herd structures.[62] Dinosaurs may have congregated in herds for defense, for migratory purposes, or to provide protection for their young. There is evidence that many types of dinosaurs, including various theropods, sauropods, ankylosaurians, ornithopods, and ceratopsians, formed aggregations of immature individuals. One example is a site in Inner Mongolia that has yielded the remains of over 20 *Sinornithomimus*, from one to seven years old. This assemblage is interpreted as a social group that was trapped in mud.[63] The interpretation of dinosaurs as gregarious has also extended to depicting carnivorous theropods as pack hunters working together to bring down large prey.[64] [65] However, this lifestyle is uncommon among the modern relatives of dinosaurs (crocodiles and other reptiles, and birds – Harris's Hawk is a well-documented exception), and the taphonomic evidence suggesting pack hunting in such theropods as *Deinonychus* and *Allosaurus* can also be interpreted as the results of fatal disputes between feeding animals, as is seen in many modern diapsid predators.[66]

Jack Horner's 1978 discovery of a *Maiasaura* ("good mother dinosaur") nesting ground in Montana demonstrated that parental care continued long after birth among the ornithopods.[67] There is also evidence that other Cretaceous-era dinosaurs, like Patagonian titanosaurian sauropods (1997 discovery), also nested in large groups.[68] The Mongolian oviraptorid *Citipati* was discovered in a chicken-like brooding position in 1993, which may mean it was covered with an insulating layer of feathers that kept the eggs warm.[69] Parental care is also implied by other finds. For example, the fossilized remains of a grouping of *Psittacosaurus* has been found, consisting of one adult and 34 juveniles; in this case, the large number of juveniles may be due to communal nesting.[70] Additionally, a

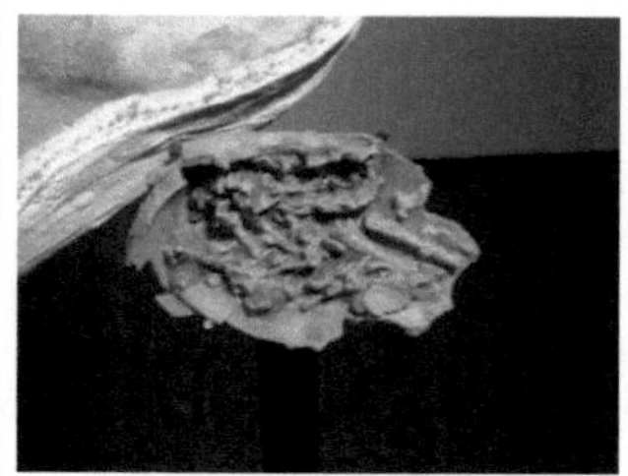

Fossilized egg of the oviraptorid *Citipati*, American Museum of Natural History

dinosaur embryo (pertaining to the prosauropod *Massospondylus*) was found without teeth, indicating that some parental care was required to feed the young dinosaur.[71] Trackways have also confirmed parental behavior among ornithopods from the Isle of Skye in northwestern Scotland.[72] Nests and eggs have been found for most major groups of dinosaurs, and it appears likely that dinosaurs communicated with their young, in a manner similar to modern birds and crocodiles.

Artist's rendering of two *Centrosaurus*, herbivorous ceratopsid dinosaurs from the late Cretaceous fauna of North America

The crests and frills of some dinosaurs, like the marginocephalians, theropods and lambeosaurines, may have been too fragile to be used for active defense, and so they were likely used for sexual or aggressive displays, though little is known about dinosaur mating and territorialism. Head wounds from bites suggest that theropods, at least, engaged in active aggressive confrontations.[73]

From a behavioral standpoint, one of the most valuable dinosaur fossils was discovered in the Gobi Desert in 1971. It included a *Velociraptor* attacking a *Protoceratops*,[74] providing evidence that dinosaurs did indeed attack each other.[75] Additional evidence for attacking live prey is the partially healed tail of an *Edmontosaurus*, a hadrosaurid dinosaur; the tail is damaged in such a way that shows the animal was bitten by a tyrannosaur but survived.[75] Cannibalism amongst some species of dinosaurs was confirmed by tooth marks found in Madagascar in 2003, involving the theropod *Majungasaurus*.[76]

Comparisons between the scleral rings of dinosaurs and modern birds and reptiles have been used to infer daily activity patterns of dinosaurs. Although it has been suggested that most dinosaurs were active during the day, these comparisons have shown that small predatory dinosaurs such as dromaeosaurids, *Juravenator*, and *Megapnosaurus* were likely nocturnal. Large and medium-sized herbivorous and omnivorous dinosaurs such as ceratopsians, sauropodomorphs, hadrosaurids, ornithomimosaurs may have been cathemeral, active during short intervals throughout the day, although the small ornithischian *Agilisaurus* was inferred to be diurnal.[77]

Based on current fossil evidence from dinosaurs such as *Oryctodromeus*, some herbivorous species seem to have led a partially fossorial (burrowing) lifestyle,[78] and some bird-like species may have been arboreal (tree climbing), most notably primitive dromaeosaurids such as *Microraptor*[79] and the enigmatic scansoriopterygids.[80] However, most dinosaurs seem to have relied on land-based locomotion. A good understanding of how dinosaurs moved on the ground is key to models of dinosaur behavior; the science of biomechanics, in particular, has provided significant insight in this area. For example, studies of the forces exerted by muscles and gravity on dinosaurs' skeletal structure have investigated how fast dinosaurs could run,[81] whether diplodocids could create sonic booms via whip-like tail snapping,[82] and whether sauropods could float.[83]

Communication and vocalization

The nature of dinosaur communication remains enigmatic, and is an active area of research. In 2008, paleontologist Phil Senter examined the evidence for vocalization in Mesozoic animal life, including dinosaurs.[84] Senter found that, contrary to popular depictions of roaring dinosaurs in motion pictures, it is likely that most dinosaurs were not capable of creating any vocalizations. To draw this conclusion, Senter studied the distribution of vocal organs in reptiles and birds. He found that vocal cords in the larynx probably evolved multiple times among reptiles, including crocodilians, which are able to produce guttural roars. Birds, on the other hand, lack a larynx. Instead, bird calls are produced by the syrinx, a vocal organ found only in birds, and which is not related to the larynx, meaning it evolved independently from the vocal organs in reptiles. The syrinx depends on the air sac system in birds to function; specifically, it requires the presence of a *clavicular air sac* near the wishbone or collar bone. This air sac leaves distinctive marks or opening on the bones, including a distinct opening in the upper arm bone (*humerus*). While many dinosaurs show evidence of extensive air sac systems, almost none possess the clavicular air sac necessary to vocalize (one exception, *Aerosteon*, probably evolved its clavicular air sac independently of birds for reasons other than vocalization).[84]

The most primitive animals with evidence of a vocalizing syrinx are the enantironithine birds. Any bird-line archosaurs more primitive than this probably did not make vocal calls. Rather, several lines of evidence suggest that dinosaurs used primarily visual communication, in the form of distinctive-looking (and possibly brightly colored) horns, frills, crests, sails and feathers. This is similar to some modern reptile groups such as lizards, in which many forms are largely silent (though like dinosaurs they possess well-developed senses of hearing) but use complex coloration and display behaviors to communicate.[84]

Also, though they may not have been able to vocalize, some dinosaurs may have used other methods of producing sound for communication. Modern animals, including reptiles and birds, use a wide variety of non-vocal sound communication, including hissing, jaw grinding or clapping, use of environment (such as splashing), and wing beating (which would have been possible in winged maniraptoran dinosaurs).[84]

Some studies have suggested that the hollow crests of the lambeosaurines may have functioned as resonance chambers used for a wide range of vocalizations.[85] [86] However, Senter (2008) noted that such chambers are also used in modern non-vocal animals to accentuate or deepen non-vocal sounds like hissing. For example, many snakes, which lack vocal cords, have resonating chambers in the skull.[84]

Physiology

A vigorous debate on the subject of temperature regulation in dinosaurs has been ongoing since the 1960s. Originally, scientists broadly disagreed as to whether dinosaurs were capable of regulating their body temperatures at all. More recently, dinosaur endothermy has become the consensus view, and debate has focused on the mechanisms of temperature regulation.

After dinosaurs were discovered, paleontologists first posited that they were ectothermic creatures: "terrible lizards" as their name suggests. This supposed cold-bloodedness was used to imply that dinosaurs were relatively slow, sluggish organisms, even though many modern reptiles

Tyrannosaurus rex skull and upper vertebral column, Palais de la Découverte, Paris

are fast and light-footed despite relying on external sources of heat to regulate their body temperature. The idea of dinosaurs as ectothermic and sluggish remained a prevalent view until Robert T. "Bob" Bakker, an early proponent of dinosaur endothermy, published an influential paper on the topic in 1968.

Modern evidence indicates that dinosaurs thrived in cooler temperate climates, and that at least some dinosaur species must have regulated their body temperature by internal biological means (perhaps aided by the animals' bulk). Evidence of endothermy in dinosaurs includes the discovery of polar dinosaurs in Australia and Antarctica (where they would have experienced a cold, dark six-month winter), the discovery of dinosaurs whose feathers may have provided regulatory insulation, and analysis of blood-vessel structures within dinosaur bone that are typical of endotherms. Skeletal structures suggest that theropods and some other dinosaurs had active lifestyles better suited to an endothermic cardiovascular system, while sauropods exhibit fewer endothermic characteristics. It is certainly possible that some dinosaurs were endothermic while others were not. Scientific debate over the specifics continues.[87]

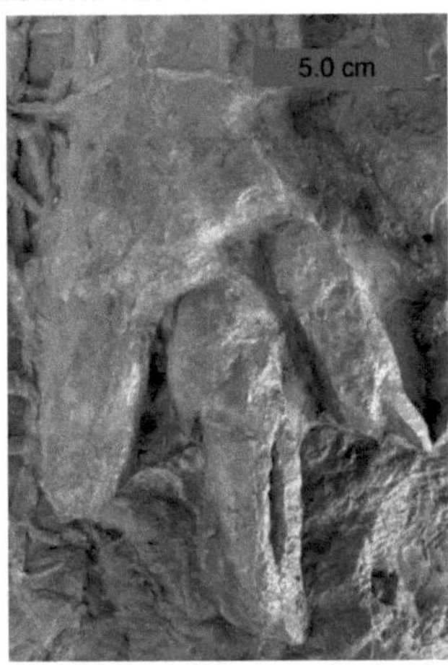

Eubrontes, a dinosaur footprint in the Lower Jurassic Moenave Formation at the St. George Dinosaur Discovery Site at Johnson Farm, southwestern Utah

Complicating the debate is the fact that warm-bloodedness can emerge based on more than one mechanism. Most discussions of dinosaur endothermy tend to compare them with average-sized birds or mammals, which expend energy to elevate body temperature above that of the environment. Small birds and mammals also possess insulation, such as fat, fur, or feathers, which slows down heat loss. However, large mammals, such as elephants, face a different problem because of their relatively small ratio of surface area to volume (Haldane's principle). This ratio compares the volume of an animal with the area of its skin: as an animal gets bigger, its surface area increases more slowly than its volume. At a certain point, the amount of heat radiated away through the skin drops below the amount of heat produced inside the body, forcing animals to use additional methods to avoid overheating. In the case of elephants, they have little hair as adults, have large ears which increase their surface area, and have behavioral adaptations as well (such as using the trunk to spray water on themselves and mud-wallowing). These behaviors increase cooling through evaporation.

Large dinosaurs would presumably have had to deal with similar issues; their body size suggest they lost heat relatively slowly to the surrounding air, and so could have been what are called inertial homeotherms, animals that are warmer than their environments through sheer size rather than through special adaptations like those of birds or mammals. However, so far this theory fails to account for the numerous dog- and goat-sized dinosaur species, or the young of larger species.

Modern computerized tomography (CT) scans of a dinosaur's chest cavity (conducted in 2000) found the apparent remnants of a four-chambered heart, much like those found in today's mammals and birds.[88] The idea is controversial within the scientific community, coming under fire for bad anatomical science[89] or simply wishful thinking.[90] The question of how this find reflects on metabolic rate and dinosaur internal anatomy may be moot, though, regardless of the object's identity: both modern crocodilians and birds, the closest living relatives of dinosaurs, have four-chambered hearts (albeit modified in crocodilians), and so dinosaurs probably had them as well.[91]

Soft tissue and DNA

One of the best examples of soft-tissue impressions in a fossil dinosaur was discovered in Petraroia, Italy. The discovery was reported in 1998, and described the specimen of a small, very young coelurosaur, *Scipionyx samniticus*. The fossil includes portions of the intestines, colon, liver, muscles, and windpipe of this immature dinosaur.[46]

In the March 2005 issue of *Science*, the paleontologist Mary Higby Schweitzer and her team announced the discovery of flexible material resembling actual soft tissue inside a 68-million-year-old *Tyrannosaurus rex* leg bone from the Hell Creek Formation in Montana. After recovery, the tissue was rehydrated by the science team.[47]

When the fossilized bone was treated over several weeks to remove mineral content from the fossilized bone-marrow cavity (a process called demineralization), Schweitzer found evidence of intact structures such as blood vessels, bone matrix, and connective tissue (bone fibers). Scrutiny under the microscope further revealed that the putative dinosaur soft tissue had retained fine structures (microstructures) even at the cellular level. The exact nature and composition of this material, and the implications of Schweitzer's discovery, are not yet clear; study and interpretation of the material is ongoing.[47]

Newer research, published in PloS One (30 July 2008), has challenged the claims that the material found is the soft tissue of *Tyrannosaurus*. Thomas Kaye of the University of Washington and his co-authors contend that what was really inside the tyrannosaur bone was slimy biofilm created by bacteria that coated the voids once occupied by blood vessels and cells.[92] The researchers found that what previously had been identified as remnants of blood cells, because of the presence of iron, were actually framboids, microscopic mineral spheres bearing iron. They found similar spheres in a variety of other fossils from various periods, including an ammonite. In the ammonite they found the spheres in a place where the iron they contain could not have had any relationship to the presence of blood.[93]

The successful extraction of ancient DNA from dinosaur fossils has been reported on two separate occasions, but, upon further inspection and peer review, neither of these reports could be confirmed.[94] However, a functional visual peptide of a theoretical dinosaur has been inferred using analytical phylogenetic reconstruction methods on gene sequences of related modern species such as reptiles and birds.[95] In addition, several proteins, including hemoglobin,[96] have putatively been detected in dinosaur fossils.[97]

Feathers and the origin of birds

The possibility that dinosaurs were the ancestors of birds was first suggested in 1868 by Thomas Henry Huxley.[98] After the work of Gerhard Heilmann in the early 20th century, the theory of birds as dinosaur descendants was abandoned in favor of the idea of their being descendants of generalized thecodonts, with the key piece of evidence being the supposed lack of clavicles in dinosaurs.[99] However, as later discoveries showed, clavicles (or a single fused wishbone, which derived from separate clavicles) were not actually absent;[14] they had been found as early as 1924 in *Oviraptor*, but misidentified as an interclavicle.[100] In the 1970s, John Ostrom revived the dinosaur–bird theory,[101] which gained momentum in the coming decades with the advent of cladistic analysis,[102] and a great increase in the discovery of small theropods and early birds.[22] Of particular note have been the fossils of the Yixian Formation, where a variety of theropods and early birds have been found, often with feathers of some type.[14] Birds share over a hundred distinct anatomical features with theropod dinosaurs, which are now generally accepted to have been their closest ancient relatives.[103] They are most closely allied with maniraptoran coelurosaurs.[14] A minority of scientists, most notably Alan Feduccia and Larry Martin, have proposed other evolutionary paths, including revised versions of Heilmann's basal archosaur proposal,[104] or that maniraptoran theropods are the ancestors of birds but themselves are not dinosaurs, only convergent with dinosaurs.[105]

Feathers

Archaeopteryx, the first good example of a "feathered dinosaur", was discovered in 1861. The initial specimen was found in the Solnhofen limestone in southern Germany, which is a *lagerstätte*, a rare and remarkable geological formation known for its superbly detailed fossils. *Archaeopteryx* is a transitional fossil, with features clearly intermediate between those of modern reptiles and birds. Brought to light just two years after Darwin's seminal *The Origin of Species*, its discovery spurred the nascent debate between proponents of evolutionary biology and creationism. This early bird is so dinosaur-like that, without a clear impression of feathers in the surrounding rock, at least one specimen was mistaken for *Compsognathus*.[106]

The famous Berlin Specimen of *Archaeopteryx lithographica*

Since the 1990s, a number of additional feathered dinosaurs have been found, providing even stronger evidence of the close relationship between dinosaurs and modern birds. Most of these specimens were unearthed in the lagerstätte of the Yixian Formation, Liaoning, northeastern China, which was part of an island continent during the Cretaceous. Though feathers have been found in only a few locations, it is possible that non-avian dinosaurs elsewhere in the world were also feathered. The lack of widespread fossil evidence for feathered non-avian dinosaurs may be because delicate features like skin and feathers are not often preserved by fossilization and thus are absent from the fossil record. To this point, protofeathers (thin, filament-like structures) are known from dinosaurs at the base of Coelurosauria, such as compsognathids like *Sinosauropteryx* and tyrannosauroids (*Dilong*),[107] but barbed feathers are known only among the coelurosaur subgroup Maniraptora, which includes oviraptorosaurs, troodontids, dromaeosaurids, and birds.[14] [108] The description of feathered dinosaurs has not been without controversy; perhaps the most vocal critics have been Alan Feduccia and Theagarten Lingham-Soliar, who have proposed that protofeathers are the result of the decomposition of collagenous fiber that underlaid the dinosaurs' integument,[109] [110] [111] and that maniraptoran dinosaurs with barbed feathers were not actually dinosaurs, but convergent with dinosaurs.[105] [110] However, their views have for the most part not been accepted by other researchers, to the point that the question of the scientific nature of Feduccia's proposals has been raised.[112]

Skeleton

Because feathers are often associated with birds, feathered dinosaurs are often touted as the missing link between birds and dinosaurs. However, the multiple skeletal features also shared by the two groups represent another important line of evidence for paleontologists. Areas of the skeleton with important similarities include the neck, pubis, wrist (semi-lunate carpal), arm and pectoral girdle, furcula (wishbone), and breast bone. Comparison of bird and dinosaur skeletons through cladistic analysis strengthens the case for the link.

Soft anatomy

Large meat-eating dinosaurs had a complex system of air sacs similar to those found in modern birds, according to an investigation which was led by Patrick O'Connor of Ohio University. The lungs of theropod dinosaurs (carnivores that walked on two legs and had bird-like feet) likely pumped air into hollow sacs in their skeletons, as is the case in birds. "What was once formally considered unique to birds was present in some form in the ancestors of birds", O'Connor said.[113] In a 2008 paper published in the online journal *PLoS ONE*, scientists described *Aerosteon riocoloradensis*, the skeleton of which supplies the strongest evidence to date of a dinosaur with a bird-like breathing system. CT-scanning revealed the evidence of air sacs within the body cavity of the *Aerosteon* skeleton.[114] [115]

Pneumatopores on the left ilium of *Aerosteon riocoloradensis*

Reproductive biology

A discovery of features in a *Tyrannosaurus rex* skeleton provided evidence of medullary bone in dinosaurs and, for the first time, allowed paleontologists to establish the sex of a dinosaur. When laying eggs, female birds grow a special type of bone between the hard outer bone and the marrow of their limbs. This medullary bone, which is rich in calcium, is used to make eggshells. The presence of endosteally derived bone tissues lining the interior marrow cavities of portions of the *Tyrannosaurus rex* specimen's hind limb suggested that *T. rex* used similar reproductive strategies, and revealed the specimen to be female.[116] Further research has found medullary bone in the theropod *Allosaurus* and the ornithopod *Tenontosaurus*. Because the line of dinosaurs that includes *Allosaurus* and *Tyrannosaurus* diverged from the line that led to *Tenontosaurus* very early in the evolution of dinosaurs, this suggests that dinosaurs in general produced medullary tissue. Medullary bone has been found in specimens of sub-adult size, which suggests that dinosaurs reached sexual maturity rather quickly for such large animals.[117]

Behavioral evidence

Fossils of the troodonts *Mei* and *Sinornithoides* demonstrate that some dinosaurs slept with their heads tucked under their arms.[118] This behavior, which may have helped to keep the head warm, is also characteristic of modern birds. Several deinonychosaur and oviraptorosaur specimens have also been found preserved on top of their nests, likely brooding in a bird-like manner.[119] The ratio between egg volume and body mass of adults among these dinosaurs suggest that the eggs were primarily brooded by the male, and that the young were highly precocial, similar to many modern ground-dwelling birds.[120]

Some dinosaurs are known to have used gizzard stones like modern birds. These stones are swallowed by animals to aid digestion and break down food and hard fibers once they enter the stomach. When found in association with fossils, gizzard stones are called gastroliths.[121]

Extinction of major groups

The discovery that birds are a type of dinosaur showed that dinosaurs in general are not, in fact, extinct as is commonly stated.[122] However, all non-avian dinosaurs as well as many groups of birds did suddenly become extinct approximately 65 million years ago. Many other groups of animals also became extinct at this time, including ammonites (nautilus-like mollusks), mosasaurs, plesiosaurs, pterosaurs, and many groups of mammals.[6] This mass extinction is known as the Cretaceous–Paleogene extinction event. The nature of the event that caused this mass extinction has been extensively studied since the 1970s; at present, several related theories are supported by paleontologists. Though the consensus is that an impact event was the primary cause of dinosaur extinction, some scientists cite other possible causes, or support the idea that a confluence of several factors was responsible for the sudden disappearance of dinosaurs from the fossil record.

At the peak of the Mesozoic, there were no polar ice caps, and sea levels are estimated to have been from 100 to 250 meters (300 to 800 ft) higher than they are today. The planet's temperature was also much more uniform, with only 25 °C (45 °F) separating average polar temperatures from those at the equator. On average, atmospheric temperatures were also much higher; the poles, for example, were 50 °C (90 °F) warmer than today.[123] [124]

The atmosphere's composition during the Mesozoic was vastly different as well. Carbon dioxide levels were up to 12 times higher than today's levels, and oxygen formed 32 to 35% of the atmosphere, as compared to 21% today. However, by the late Cretaceous, the environment was changing dramatically. Volcanic activity was decreasing, which led to a cooling trend as levels of atmospheric carbon dioxide dropped. Oxygen levels in the atmosphere also started to fluctuate and would ultimately fall considerably. Some scientists hypothesize that climate change, combined with lower oxygen levels, might have led directly to the demise of many species. If the dinosaurs had respiratory systems similar to those commonly found in modern birds, it may have been particularly difficult for them to cope with reduced respiratory efficiency, given the enormous oxygen demands of their very large bodies.[6]

Impact event

The asteroid collision theory, which was brought to wide attention in 1980 by Walter Alvarez and colleagues, links the extinction event at the end of the Cretaceous period to a bolide impact approximately 65.5 million years ago. Alvarez *et al.* proposed that a sudden increase in iridium levels, recorded around the world in the period's rock stratum, was direct evidence of the impact.[125] The bulk of the evidence now suggests that a bolide 5 to 15 kilometers (3 to 9 mi) wide hit in the vicinity of the Yucatán Peninsula (in southeastern Mexico), creating the approximately 180 km (110 mi) Chicxulub Crater and triggering the mass extinction.[126] [127] Scientists are not certain whether dinosaurs were thriving or declining before the impact event. Some scientists propose that the meteorite caused a long and unnatural drop in Earth's atmospheric temperature, while others claim that it would have instead created an unusual heat wave. The consensus among scientists who support this theory is that the impact caused extinctions both directly (by heat from the meteorite impact) and also

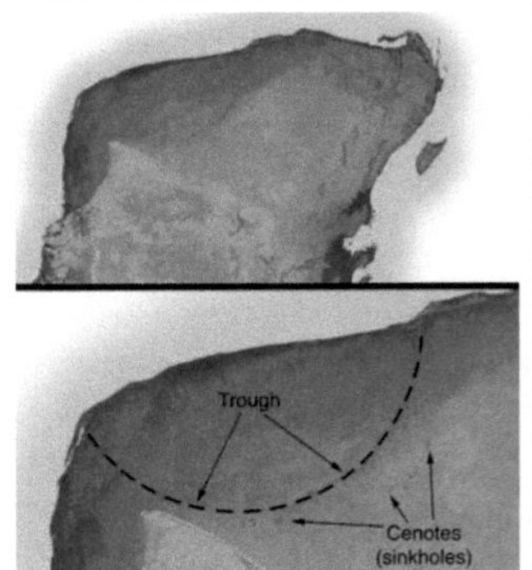

The Chicxulub Crater at the tip of the Yucatán Peninsula; the impactor that formed this crater may have caused the dinosaur extinction.

indirectly (via a worldwide cooling brought about when matter ejected from the impact crater reflected thermal radiation from the sun). Although the speed of extinction cannot be deduced from the fossil record alone, various models suggest that the extinction was extremely rapid, being down to hours rather than years.[128]

In September 2007, U.S. researchers led by William Bottke of the Southwest Research Institute in Boulder, Colorado, and Czech scientists used computer simulations to identify the probable source of the Chicxulub impact. They calculated a 90% probability that a giant asteroid named Baptistina, approximately 160 km (99 mi) in diameter, orbiting in the asteroid belt which lies between Mars and Jupiter, was struck by a smaller unnamed asteroid about 55 km (35 mi) in diameter about 160 million years ago. The impact shattered Baptistina, creating a cluster which still exists today as the Baptistina family. Calculations indicate that some of the fragments were sent hurtling into earth-crossing orbits, one of which was the 10 km (6.2 mi) wide meteorite which struck Mexico's Yucatan peninsula 65 million years ago, creating the Chicxulub crater.[129] In 2011, new data from the Wide-field Infrared Survey Explorer revised the date of the collision which created the Baptistina family to about 80 million years ago. This makes an asteroid from this family highly improbable to be the asteroid that created the Chicxulub Crater, as typically the process of resonance and collision of an asteroid takes many tens of millions of years.[130]

A similar but more controversial explanation proposes that "passages of the [hypothetical] solar companion star Nemesis through the Oort comet cloud would trigger comet showers."[131] One or more of these comets then collided with the Earth at approximately the same time, causing the worldwide extinction. As with the impact of a single asteroid, the end result of this comet bombardment would have been a sudden drop in global temperatures, followed by a protracted cool period.[131]

Deccan Traps

Before 2000, arguments that the Deccan Traps flood basalts caused the extinction were usually linked to the view that the extinction was gradual, as the flood basalt events were thought to have started around 68 million years ago and lasted for over 2 million years. However, there is evidence that two thirds of the Deccan Traps were created in only 1 million years about 65.5 million years ago, and so these eruptions would have caused a fairly rapid extinction, possibly over a period of thousands of years, but still longer than would be expected from a single impact event.[132] [133]

The Deccan Traps could have caused extinction through several mechanisms, including the release into the air of dust and sulphuric aerosols, which might have blocked sunlight and thereby reduced photosynthesis in plants. In addition, Deccan Trap volcanism might have resulted in carbon dioxide emissions, which would have increased the greenhouse effect when the dust and aerosols cleared from the atmosphere.[133] Before the mass extinction of the dinosaurs, the release of volcanic gases during the formation of the Deccan Traps "contributed to an apparently massive global warming. Some data point to an average rise in temperature of 8 °C (14 °F) in the last half million years before the impact [at Chicxulub]."[132] [133]

In the years when the Deccan Traps theory was linked to a slower extinction, Luis Alvarez (who died in 1988) replied that paleontologists were being misled by sparse data. While his assertion was not initially well-received, later intensive field studies of fossil beds lent weight to his claim. Eventually, most paleontologists began to accept the idea that the mass extinctions at the end of the Cretaceous were largely or at least partly due to a massive Earth impact. However, even Walter Alvarez has acknowledged that there were other major changes on Earth even before the impact, such as a drop in sea level and massive volcanic eruptions that produced the Indian Deccan Traps, and these may have contributed to the extinctions.[134]

Failure to adapt to changing conditions

Lloyd *et al.* (2008) noted that, in the Mid Cretaceous, the flowering, angiosperm plants became a major part of terrestrial ecosystems, which had previously been dominated by gymnosperms such as conifers. Dinosaur coprolite—fossilized dung—indicate that, while some ate angiosperms, most herbivorous dinosaurs ate mainly gymnosperms. Statistical analysis by Lloyd *et al.* concluded that, contrary to earlier studies, dinosaurs did not diversify very much in the Late Cretaceous. Lloyd *et al.* suggested that dinosaurs' failure to diversify as ecosystems were changing doomed them to extinction.[135]

Possible Paleocene survivors

Non-avian dinosaur remains are occasionally found above the K–T boundary. In 2001, paleontologists Zielinski and Budahn reported the discovery of a single hadrosaur leg-bone fossil in the San Juan Basin, New Mexico, and described it as evidence of Paleocene dinosaurs. The formation in which the bone was discovered has been dated to the early Paleocene epoch, approximately 64.5 million years ago. If the bone was not re-deposited into that stratum by weathering action, it would provide evidence that some dinosaur populations may have survived at least a half million years into the Cenozoic Era.[136] Other evidence includes the finding of dinosaur remains in the Hell Creek Formation up to 1.3 meters (51 in) above (40000 years later than) the K–T boundary. Similar reports have come from other parts of the world, including China.[137] Many scientists, however, dismissed the supposed Paleocene dinosaurs as re-worked, that is, washed out of their original locations and then re-buried in much later sediments.[138] [139] However, direct dating of the bones themselves has supported the later date, with U–Pb dating methods resulting in a precise age of 64.8 ± 0.9 million years ago.[140] If correct, the presence of a handful of dinosaurs in the early Paleocene would not change the underlying facts of the extinction.[138]

History of discovery

Dinosaur fossils have been known for millennia, although their true nature was not recognized. The Chinese, whose modern word for dinosaur is *konglong* (, or "terrible dragon"), considered them to be dragon bones and documented them as such. For example, *Hua Yang Guo Zhi*, a book written by Zhang Qu during the Western Jin Dynasty, reported the discovery of dragon bones at Wucheng in Sichuan Province.[141] Villagers in central China have long unearthed fossilized "dragon bones" for use in traditional medicines, a practice that continues today.[142] In Europe, dinosaur fossils were generally believed to be the remains of giants and other creatures killed by the Great Flood.

Scholarly descriptions of what would now be recognized as dinosaur bones first appeared in the late 17th century in England. Part of a bone, now known to have been the femur of a *Megalosaurus*,[143] was recovered from a limestone quarry at Cornwell near Chipping Norton, Oxfordshire, England, in 1676. The fragment was sent to Robert Plot, Professor of Chemistry at the University of Oxford and first curator of the Ashmolean Museum, who published a description in his *Natural History of Oxfordshire* in 1677. He correctly identified the bone as the lower extremity of the femur of a large animal, and recognized that it was too large to belong to any known species. He therefore concluded it to be the thigh bone of a giant human similar to those mentioned in

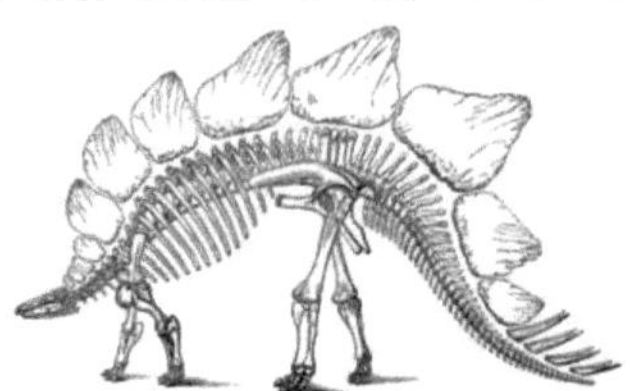

Marsh's 1896 illustration of the bones of *Stegosaurus*, a dinosaur he described and named in 1877.

the Bible. In 1699, Edward Lhuyd, a friend of Sir Isaac Newton, was responsible for the first published scientific treatment of what would now be recognized as a dinosaur when he described and named a sauropod tooth, "Rutellum implicatum",[144] [145] that had been found in Caswell, near Witney, Oxfordshire.[146]

William Buckland

Between 1815 and 1824, the Rev William Buckland, a professor of geology at Oxford University, collected more fossilized bones of *Megalosaurus* and became the first person to describe a dinosaur in a scientific journal.[143] [147] The second dinosaur genus to be identified, *Iguanodon*, was discovered in 1822 by Mary Ann Mantell – the wife of English geologist Gideon Mantell. Gideon Mantell recognized similarities between his fossils [148] and the bones of modern iguanas. He published his findings in 1825.[149] [150]

The study of these "great fossil lizards" soon became of great interest to European and American scientists, and in 1842 the English paleontologist Richard Owen coined the term "dinosaur". He recognized that the remains that had been found so far, *Iguanodon*, *Megalosaurus* and *Hylaeosaurus*, shared a number of distinctive features, and so decided to present them as a distinct taxonomic group. With the backing of Prince Albert of Saxe-Coburg-Gotha, the husband of Queen Victoria, Owen established the Natural History Museum in South Kensington, London, to display the national collection of dinosaur fossils and other biological and geological exhibits.

In 1858, the first known American dinosaur was discovered, in marl pits in the small town of Haddonfield, New Jersey (although fossils had been found before, their nature had not been correctly discerned). The creature was named *Hadrosaurus foulkii*. It was an extremely important find: *Hadrosaurus* was one of the first nearly complete dinosaur skeletons found (the first was in 1834, in Maidstone, Kent, England), and it was clearly a bipedal creature. This was a revolutionary discovery as, until that point, most scientists had believed dinosaurs walked on four feet, like other lizards. Foulke's discoveries sparked a wave of dinosaur mania in the United States.

Othniel Charles Marsh,
19th century photograph

Dinosaur mania was exemplified by the fierce rivalry between Edward Drinker Cope and Othniel Charles Marsh, both of whom raced to be the first to find new dinosaurs in what came to be known as the Bone Wars. The feud probably originated when Marsh publicly pointed out that Cope's reconstruction of an *Elasmosaurus* skeleton was flawed: Cope had inadvertently placed the plesiosaur's head at what should have been the animal's tail end. The fight between the two scientists lasted for over 30 years, ending in 1897 when Cope died after spending his entire fortune on the dinosaur hunt. Marsh 'won' the contest primarily because he was better funded through a relationship with the US Geological Survey. Unfortunately, many valuable dinosaur specimens were damaged or destroyed due to the pair's rough methods: for example, their diggers often used dynamite to unearth bones (a method modern paleontologists would find appalling). Despite their unrefined methods, the contributions of Cope and Marsh to paleontology were vast: Marsh unearthed 86 new species of dinosaur and Cope discovered 56, a total of 142 new species. Cope's collection is now at the American Museum of Natural History in New York, while Marsh's is on display at the Peabody Museum of Natural History at Yale University.[151]

Edward Drinker Cope,
19th century photograph

After 1897, the search for dinosaur fossils extended to every continent, including Antarctica. The first Antarctic dinosaur to be discovered, the ankylosaurid *Antarctopelta oliveroi*, was found on Ross Island in 1986, although it was 1994 before an Antarctic species, the theropod *Cryolophosaurus ellioti*, was formally named and described in a scientific journal.

Current dinosaur "hot spots" include southern South America (especially Argentina) and China. China in particular has produced many exceptional feathered dinosaur specimens due to the unique geology of its dinosaur beds, as well as an ancient arid climate particularly conducive to fossilization.

The "dinosaur renaissance"

The field of dinosaur research has enjoyed a surge in activity that began in the 1970s and is ongoing. This was triggered, in part, by John Ostrom's discovery of *Deinonychus*, an active predator that may have been warm-blooded, in marked contrast to the then-prevailing image of dinosaurs as sluggish and cold-blooded. Vertebrate paleontology has become a global science. Major new dinosaur discoveries have been made by paleontologists working in previously unexploited regions, including India, South America, Madagascar, Antarctica, and most significantly China (the amazingly well-preserved feathered dinosaurs in China have further consolidated the link between dinosaurs and their conjectured living descendants, modern birds). The widespread application of cladistics, which rigorously analyzes the relationships between biological organisms, has also proved tremendously useful in classifying dinosaurs. Cladistic analysis, among other modern techniques, helps to compensate for an often incomplete and fragmentary fossil record.

Cultural depictions

By human standards, dinosaurs were creatures of fantastic appearance and often enormous size. As such, they have captured the popular imagination and become an enduring part of human culture. Entry of the word "dinosaur" into the common vernacular reflects the animals' cultural importance: in English, "dinosaur" is commonly used to describe anything that is impractically large, obsolete, or bound for extinction.[7]

Public enthusiasm for dinosaurs first developed in Victorian England, where in 1854, three decades after the first scientific descriptions of dinosaur remains, the famous dinosaur sculptures were unveiled in London's Crystal Palace Park. The Crystal Palace dinosaurs proved so popular that a strong market in smaller replicas soon developed. In subsequent decades, dinosaur exhibits opened at parks and museums around the world, ensuring that successive generations would be introduced to the animals in an immersive and exciting way.[152] Dinosaurs' enduring popularity, in its turn, has resulted in significant public funding for dinosaur science, and has frequently spurred new discoveries. In the United States, for example, the competition between museums for public attention led directly to the Bone Wars of the 1880s and 1890s, during which a pair of feuding paleontologists made enormous scientific contributions.[153]

The popular preoccupation with dinosaurs has ensured their appearance in literature, film and other media. Beginning in 1852 with a passing mention in Charles Dickens' *Bleak House*,[154] dinosaurs have been featured in large numbers of fictional works. Sir Arthur Conan Doyle's 1912 book *The Lost World*, the iconic 1933 film *King Kong*, 1954's *Godzilla* and its many sequels, the best-selling 1990 novel *Jurassic Park* by Michael Crichton and its 1993 film adaptation are just a few notable examples of dinosaur appearances in fiction. Authors of general-interest non-fiction works about dinosaurs, including some prominent paleontologists, have often sought to use the animals as a way to educate readers about science in general. Dinosaurs are ubiquitous in advertising; numerous companies have referenced dinosaurs in printed or televised advertisements, either in order to sell their own products or in order to characterize their rivals as slow-moving, dim-witted or obsolete.[155]

Notes and references

[1] Gauthier, Jacques; de Querioz, Kevin (2001). "Feathered dinosaurs, flying dinosaurs, crown dinosaurs, and the name 'Aves'." (http:// vertebrates.si.edu/herps/herps_pdfs/deQueiroz_pdfs/2001gaudeqost.pdf) (PDF). *New Perspectives on the Origin and Early Evolution of Birds: Proceedings of the International Symposium in Honor of John H. Ostrom*. Peabody Museum of Natural History, Yale University. ISBN 0-912532-57-2. . Retrieved 2009-09-22.

[2] Zhou, Z. (2004). "The origin and early evolution of birds: discoveries, disputes, and perspectives from fossil evidence". *Naturwissenchaften* **91** (10): 455–471. Bibcode 2004NW.....91..455Z. doi:10.1007/s00114-004-0570-4.

[3] Alfaro, M.E., F. Santini, C. Brock, H. Alamillo, A. Dornburg. D.L. Rabosky, G. Carnevale, and L.J. Harmon (2009). "Nine exceptional radiations plus high turnover explain species diversity in jawed vertebrates". *Proceedings of the National Academy of Sciences USA* **106** (32): 13410–13414. Bibcode 2009PNAS..10613410A. doi:10.1073/pnas.0811087106. PMC 2715324. PMID 19633192.

[4] Wang, S.C., and Dodson, P. (2006). "Estimating the Diversity of Dinosaurs". *Proceedings of the National Academy of Sciences USA* **103** (37): 13601–13605. Bibcode 2006PNAS..10313601W. doi:10.1073/pnas.0606028103. PMC 1564218. PMID 16954187.

[5] Amos J (2008-09-17). "Will the real dinosaurs stand up?" (http://news.bbc.co.uk/2/hi/science/nature/7620621.stm). *BBC News*. . Retrieved 2011-03-23.

[6] MacLeod, N, Rawson, PF, Forey, PL, Banner, FT, Boudagher-Fadel, MK, Bown, PR, Burnett, JA, Chambers, P, Culver, S, Evans, SE, Jeffery, C, Kaminski, MA, Lord, AR, Milner, AC, Milner, AR, Morris, N, Owen, E, Rosen, BR, Smith, AB, Taylor, PD, Urquhart, E & Young, JR (1997). "The Cretaceous–Tertiary biotic transition" (http://findarticles.com/p/articles/mi_qa3721/is_199703/ai_n8738406/ print). *Journal of the Geological Society* **154** (2): 265–292. doi:10.1144/gsjgs.154.2.0265. .

[7] "Dinosaur – Definition and More" (http://www.m-w.com/dictionary/dinosaur). Merriam-Webster Dictionary. . Retrieved 2011-05-06.

[8] Owen, R (1842)). *Report on British Fossil Reptiles." Part II*. Report of the Eleventh Meeting of the British Association for the Advancement of Science; Held at Plymouth in July 1841 (http://books.google.com/books?id=dy5LAAAAYAAJ&pg=PA60&f=false#v=onepage&q& f=false). London: John Murray. pp. 60–204. .

[9] "Liddell–Scott–Jones Lexicon of Classical Greek" (http://www.perseus.tufts.edu/cgi-bin/lexindex?lookup=deino/s&lang=greek& doc=Perseus:text:1999.01.0169&formentry=0). . Retrieved 2008-08-05.

[10] Farlow, J.O., and Brett-Surman, M.K. (1997). "Preface". In Farlow, J.O., and Brett-Surman, M.K. (eds.). *The Complete Dinosaur*. Indiana University Press. pp. ix–xi. ISBN 0-253-33349-0.

[11] Benton, Michael J. (2004). "Origin and relationships of Dinosauria". In Weishampel, David B.; Dodson, Peter; and Osmólska, Halszka (eds.). *The Dinosauria* (2nd ed.). Berkeley: University of California Press. pp. 7–19. ISBN 0-520-24209-2.

[12] Olshevsky, G. (2000). "An annotated checklist of dinosaur species by continent". *Mesozoic Meanderings* **3**: 1–157.

[13] Sereno, P. (2005). "The logical basis of phylogenetic taxonomy". *Systematic Biology* **54** (4): 595–619.

[14] Padian K (2004). "Basal avialae". In Weishampel DB, Dodson P, Osmólska H. *The Dinosauria (2d edition)*. University of California Press. pp. 210–231. ISBN 0-520-24209-2.

[15] Glut, Donald F. (1997). *Dinosaurs: The Encyclopedia*. Jefferson, North Carolina: McFarland & Co. p. 40. ISBN 0-89950-917-7.

[16] Lambert, David; and the Diagram Group (1990). *The Dinosaur Data Book*. New York: Avon Books. p. 288. ISBN 0-380-75896-2.

[17] Morales, Michael (1997). "Nondinosaurian vertebrates of the Mesozoic". In Farlow JO, Brett-Surman MK. *The Complete Dinosaur*. Bloomington: Indiana University Press. pp. 607–624. ISBN 0-253-33349-0.

[18] Hu Yaoming; Meng, J; Wang, Y; Li, C (2005). "Large Mesozoic mammals fed on dinosaurs". *Nature* **433** (7022): 149–152. Bibcode 2005Natur.433..149H. doi:10.1038/nature03102. PMID 15650737.

[19] Russell, Dale A. (1995). "China and the lost worlds of the dinosaurian era". *Historical Biology* **10**: 3–12. doi:10.1080/10292389509380510.

[20] Amiot, R.; Buffetaut, E.; Lécuyer, C.; Wang, X.; Boudad, L.; Ding, Z.; Fourel, F.; Hutt, S.; Martineau, F.; Medeiros, A.; Mo, J.; Simon, L.; Suteethorn, V.; Sweetman, S.; Tong, H.; Zhang, F.; and Zhou, Z. (2010). "Oxygen isotope evidence for semi-aquatic habits among spinosaurid theropods". *Geology* **38** (2): 139–142. doi:10.1130/G30402.1.

[21] Nesbitt S.J. (2011). "The early evolution of archosaurs : relationships and the origin of major clades". *Bulletin of the American Museum of Natural History* **352**: 1–292. doi:10.1206/352.1.

[22] Holtz, Jr., T.R. (2000). "Classification and evolution of the dinosaur groups". In Paul, G.S.. *The Scientific American Book of Dinosaurs*. St. Martin's Press. pp. 140–168. ISBN 0-312-26226-4.

[23] Langer, M.C., Abdala, F., Richter, M., and Benton, M.J. (1999). "A sauropodomorph dinosaur from the Upper Triassic (Carnian) of southern Brazil". *Comptes Rendus de l'Academie des Sciences, Paris: Sciences de la terre et des planètes* **329**: 511–517.

[24] Nesbitt, Sterling J.; Irmis, Randall B.; Parker, William G. (2007). "A critical re-evaluation of the Late Triassic dinosaur taxa of North America". *Journal of Systematic Palaeontology* **5** (2): 209–243. doi:10.1017/S1477201907002040.

[25] This was recognized not later than 1909: "Dr. Holland and the Sprawling Sauropods" (http://www.hmnh.org/library/diplodocus/holland1910.html). . The arguments and many of the images are also presented in Desmond, A. (1976). *Hot Blooded Dinosaurs*. DoubleDay. ISBN 0-385-27063-1.

[26] Benton, M.J. (2004). *Vertebrate Paleontology*. Blackwell Publishers. xii–452. ISBN 0-632-05614-2.

[27] Cowen, Richard (2004). "Dinosaurs". *History of Life* (4th ed.). Blackwell Publishing. pp. 151–175. ISBN 1-4051-1756-7. OCLC 53970577.

[28] Kubo, T.; Benton, Michael J. (2007). "Evolution of hindlimb posture in archosaurs: limb stresses in extinct vertebrates". *Palaeontology* **50** (6): 1519–1529. doi:10.1111/j.1475-4983.2007.00723.x.

[29] Seeley, H.G. (1887). "On the classification of the fossil animals commonly named Dinosauria". *Proc R Soc London* (Royal Society) **43**: 165–171. Bibcode 1887RSPS...43..165S. doi:10.1098/rspl.1887.0117.

[30] Romer, A.S. (1956). *Osteology of the Reptiles*. University of Chicago. ISBN 0-89464-985-X.

[31] Ostrom, J.H. (1980). "The evidence of endothermy in dinosaurs". In Thomas, R.D.K. and Olson, E.C.. *A cold look at the warm-blooded dinosaurs*. Boulder, CO: American Association for the Advancement of Science. pp. 82–105.

[32] Bakker, R. T., and Galton, P (1974). "Dinosaur monophyly and a new class of vertebrates". *Nature* **248** (5444): 168–172. Bibcode 1974Natur.248..168B. doi:10.1038/248168a0.

[33] Kump LR, Pavlov A & Arthur MA (2005). "Massive release of hydrogen sulfide to the surface ocean and atmosphere during intervals of oceanic anoxia". *Geology* **33** (5): 397–400. Bibcode 2005Geo....33..397K. doi:10.1130/G21295.1.

[34] Tanner LH, Lucas SG & Chapman MG (2004). "Assessing the record and causes of Late Triassic extinctions" (http://web.archive.org/web/20071025225841/http://nmnaturalhistory.org/pdf_files/TJB.pdf) (PDF). *Earth-Science Reviews* **65** (1–2): 103–139. Bibcode 2004ESRv...65..103T. doi:10.1016/S0012-8252(03)00082-5. Archived from the original (http://nmnaturalhistory.org/pdf_files/TJB.pdf) on October 25, 2007. . Retrieved 2007-10-22.

[35] Sereno PC (1999). "The evolution of dinosaurs". *Science* **284** (5423): 2137–2147. doi:10.1126/science.284.5423.2137. PMID 10381873.

[36] Sereno, P.C.; Forster, Catherine A.; Rogers, Raymond R.; Monetta, Alfredo M. (1993). "Primitive dinosaur skeleton from Argentina and the early evolution of Dinosauria". *Nature* **361** (6407): 64–66. Bibcode 1993Natur.361...64S. doi:10.1038/361064a0.

[37] Brusatte, S.L.; Benton, MJ; Ruta, M; Lloyd, GT (2008). "Superiority, competition, and opportunism in the evolutionary radiation of dinosaurs". *Science* **321** (5895): 1485–1488. Bibcode 2008Sci...321.1485B. doi:10.1126/science.1161833. PMID 18787166.

[38] Paul, G.S. (1988). *Predatory Dinosaurs of the World*. New York: Simon and Schuster. pp. 248–250. ISBN 0-671-61946-2.

[39] Clark J.M., Maryanska T., Barsbold R (2004). "Therizinosauroidea". In Weishampel DB, Dodson P, Osmólska H. *The Dinosauria (2d edition)*. University of California Press. pp. 151–164. ISBN 0-520-24209-2.

[40] Norell MA, Makovicky PJ (2004). "Dromaeosauridae". In Weishampel DB, Dodson P, Osmólska H. *The Dinosauria (2d edition)*. University of California Press. pp. 196–210. ISBN 0-520-24209-2.

[41] Holtz, Thomas R., Jr.; Chapman, Ralph E.; and Lamanna, Matthew C. (2004). "Mesozoic biogeography of Dinosauria". In Weishampel, David B.; Dodson, Peter; and Osmólska, Halszka (eds.). *The Dinosauria* (2nd ed.). Berkeley: University of California Press. pp. 627–642. ISBN 0-520-24209-2.

[42] Fastovsky, David E.; and Smith, Joshua B. (2004). "Dinosaur paleoecology". In Weishampel, David B.; Dodson, Peter; and Osmólska, Halszka. *The Dinosauria* (2nd ed.). Berkeley: University of California Press. pp. 614–626. ISBN 0-520-24209-2.

[43] Sereno, P.C.; Wilson, JA; Witmer, LM; Whitlock, JA; Maga, A; Ide, O; Rowe, TA; Kemp, Tom (2007). "Structural extremes in a Cretaceous dinosaur". *PLoS ONE* **2** (11): e1230. Bibcode 2007PLoSO...2.1230S. doi:10.1371/journal.pone.0001230. PMC 2077925. PMID 18030355.

[44] Prasad, V.; Strömberg, CA; Alimohammadian, H; Sahni, A (2005). "Dinosaur coprolites and the early evolution of grasses and grazers". *Science* **310** (5751): 1170–1180. Bibcode 2005Sci...310.1177P. doi:10.1126/science.1118806. PMID 16293759.

[45] Archibald, J. David; and Fastovsky, David E. (2004). "Dinosaur Extinction". In Weishampel, David B.; Dodson, Peter; and Osmólska, Halszka (eds.). *The Dinosauria* (2nd ed.). Berkeley: University of California Press. pp. 672–684. ISBN 0-520-24209-2.

[46] Dal Sasso, C. and Signore, M. (1998). "Exceptional soft-tissue preservation in a theropod dinosaur from Italy". *Nature* **292** (6674): 383–387. Bibcode 1998Natur.392..383D. doi:10.1038/32884.

[47] Schweitzer, M.H., Wittmeyer, J.L. and Horner, J.R. (2005). "Soft-Tissue Vessels and Cellular Preservation in Tyrannosaurus rex". *Science* **307** (5717): 1952–1955. Bibcode 2005Sci...307.1952S. doi:10.1126/science.1108397. PMID 15790853.

[48] Carpenter, Kenneth (2006). "Biggest of the big: a critical re-evaluation of the mega-sauropod *Amphicoelias fragillimus*". In Foster, John R.; and Lucas, Spencer G. (eds.). *Paleontology and Geology of the Upper Jurassic Morrison Formation*. New Mexico Museum of Natural History and Science Bulletin **36**. Albuquerque: New Mexico Museum of Natural History and Science. pp. 131–138.

[49] Farlow JA (1993). "On the rareness of big, fierce animals: speculations about the body sizes, population densities, and geographic ranges of predatory mammals and large, carnivorous dinosaurs". In Dodson, Peter; and Gingerich, Philip. *Functional Morphology and Evolution*. American Journal of Science, Special Volume **293-A**. pp. 167–199.

[50] Peczkis, J. (1994). "Implications of body-mass estimates for dinosaurs". *Journal of Vertebrate Paleontology* **14** (4): 520–33. doi:10.1080/02724634.1995.10011575.

[51] "Anatomy and evolution" (http://paleobiology.si.edu/dinosaurs/info/everything/evo_1.html). National Museum of Natural History. . Retrieved 2007-11-21.

[52] Colbert, Edwin Harris (1971). *Men and dinosaurs: the search in field and laboratory*. Harmondsworth [Eng.]: Penguin. ISBN 0-14-021288-4.

[53] Lovelace, David M. (2007). "Morphology of a specimen of *Supersaurus* (Dinosauria, Sauropoda) from the Morrison Formation of Wyoming, and a re-evaluation of diplodocid phylogeny". *Arquivos do Museu Nacional* **65** (4): 527–544.

[54] dal Sasso C, Maganuco S, Buffetaut E, Mendez MA (2006). "New information on the skull of the enigmatic theropod *Spinosaurus*, with remarks on its sizes and affinities" (http://www.reocities.com/Athens/bridge/4602/spinoskull.pdf) (PDF). *Journal of Vertebrate Paleontology* **25** (4): 888–896. doi:10.1671/0272-4634(2005)025[0888:NIOTSO]2.0.CO;2. . Retrieved 2011-05-05.

[55] Therrien, F.; and Henderson, D.M. (2007). "My theropod is bigger than yours...or not: estimating body size from skull length in theropods". *Journal of Vertebrate Paleontology* **27** (1): 108–115. doi:10.1671/0272-4634(2007)27[108:MTIBTY]2.0.CO;2.

[56] Zhang F, Zhou Z, Xu X, Wang X, Sullivan C (2008). "A bizarre Jurassic maniraptoran from China with elongate ribbon-like feathers". *Nature* **455** (7216): 1105–1108. Bibcode 2008Natur.455.1105Z. doi:10.1038/nature07447. PMID 18948955.

[57] Xu X, Zhao Q, Norell M, Sullivan C, Hone D, Erickson G, Wang XL, Han FL, Guo Y. "A new feathered maniraptoran dinosaur fossil that fills a morphological gap in avian origin". *Chinese Science Bulletin* **54** (3): 430–435. doi:10.1007/s11434-009-0009-6.

[58] Rey LV, Holtz, Jr TR (2007). *Dinosaurs: the most complete, up-to-date encyclopedia for dinosaur lovers of all ages*. New York: Random House. ISBN 0-375-82419-7.

[59] Butler, R.J.; Zhao, Q. (2009). "The small-bodied ornithischian dinosaurs *Micropachycephalosaurus hongtuyanensis* and *Wannanosaurus yansiensis* from the Late Cretaceous of China". *Cretaceous Research* **30** (1): 63–77. doi:10.1016/j.cretres.2008.03.002.

[60] Yans J, Dejax J, Pons D, Dupuis C & Taquet P (2005). "Implications paléontologiques et géodynamiques de la datation palynologique des sédiments à faciès wealdien de Bernissart (bassin de Mons, Belgique)" (in French). *Comptes Rendus Palevol* **4** (1–2): 135–150. doi:10.1016/j.crpv.2004.12.003.

[61] Day, J.J.; Upchurch, P; Norman, DB; Gale, AS; Powell, HP (2002). "Sauropod trackways, evolution, and behavior". *Science* **296** (5573): 1659. doi:10.1126/science.1070167. PMID 12040187.

[62] Wright, Joanna L. (2005). "Steps in understanding sauropod biology". In Curry Rogers, Kristina A.; and Wilson, Jeffrey A.. *The Sauropods: Evolution and Paleobiology*. Berkeley: University of California Press. pp. 252–284. ISBN 0-520-24623-3.

[63] Varricchio, D.J.; Sereno, Paul C.; Xijin, Zhao; Lin, Tan; Wilson, Jeffery A.; Lyon, Gabrielle H. (2008). "Mud-trapped herd captures evidence of distinctive dinosaur sociality" (http://www.app.pan.pl/archive/published/app53/APP53-567.pdf) (PDF). *Acta Palaeontologica Polonica* **53** (4): 567–578. doi:10.4202/app.2008.0402. . Retrieved 2011-05-06.

[64] Lessem, Don; and Glut, Donald F. (1993). "*Allosaurus*". *The Dinosaur Society's Dinosaur Encyclopedia*. Random House. pp. 19–20. ISBN 0-679-41770-2.

[65] Maxwell, W. D.; Ostrom, John (1995). "Taphonomy and paleobiological implications of *Tenontosaurus–Deinonychus* associations". *Journal of Vertebrate Paleontology* **15** (4): 707–712. doi:10.1080/02724634.1995.10011256.(abstract (http://www.vertpaleo.org/publications/jvp/15-707-712.cfm))

[66] Roach, Brian T.; Brinkman, Daniel L. (2007). "A reevaluation of cooperative pack hunting and gregariousness in *Deinonychus antirrhopus* and other nonavian theropod dinosaurs". *Bulletin of the Peabody Museum of Natural History* **48** (1): 103–138. doi:10.3374/0079-032X(2007)48[103:AROCPH]2.0.CO;2.

[67] Horner, J.R.; Makela, Robert (1979). "Nest of juveniles provides evidence of family structure among dinosaurs". *Nature* **282** (5736): 296–298. Bibcode 1979Natur.282..296H. doi:10.1038/282296a0.

[68] Chiappe, Luis M.; Jackson, Frankie; Coria, Rodolfo A.; and Dingus, Lowell (2005). "Nesting titanosaurs from Auca Mahuevo and adjacent sites". In Curry Rogers, Kristina A.; and Wilson, Jeffrey A.. *The Sauropods: Evolution and Paleobiology*. Berkeley: University of California Press. pp. 285–302. ISBN 0-520-24623-3.

[69] "Discovering Dinosaur Behavior: 1960–present view" (http://search.eb.com/dinosaurs/dinosaurs/BRa.html+). Encyclopedia Brittanica. . Retrieved 2011-05-05.

[70] Meng Qingjin; Liu Jinyuan; Varricchio, David J.; Huang, Timothy; and Gao Chunling (2004). "Parental care in an ornithischian dinosaur". *Nature* **431** (7005): 145–146. Bibcode 2004Natur.431..145M. doi:10.1038/431145a. PMID 15356619.

[71] Reisz RR, Scott, D Sues, H-D, Evans, DC & Raath, MA (2005). "Embryos of an Early Jurassic prosauropod dinosaur and their evolutionary significance". *Science* **309** (5735): 761–764. Bibcode 2005Sci...309..761R. doi:10.1126/science.1114942. PMID 16051793.

[72] Clark NDL, Booth P, Booth CL, Ross DA (2004). "Dinosaur footprints from the Duntulm Formation (Bathonian, Jurassic) of the Isle of Skye" (http://testservice-eprints.gla.ac.uk/4496/1/4496.pdf) (PDF). *Scottish Journal of Geology* **40** (1): 13–21. . Retrieved 2011-05-05.

[73] Tanke, Darren H. (1998). "Head-biting behavior in theropod dinosaurs: paleopathological evidence" (http://www.mnhn.ul.pt/geologia/gaia/12.pdf) (PDF). *Gaia* (15): 167–184. ISSN 0871-5424. .

[74] "The Fighting Dinosaurs" (http://www.amnh.org/exhibitions/fightingdinos/ex-fd.html). American Museum of Natural History. . Retrieved 2007-12-05.

[75] Carpenter, K. (1998). "Evidence of predatory behavior by theropod dinosaurs" (http://vertpaleo.org/publications/jvp/15-576-591.cfm). *Gaia* **15**: 135–144. . Retrieved 2007-12-05.

[76] Rogers, Raymond R.; Krause, DW; Curry Rogers, K (2007). "Cannibalism in the Madagascan dinosaur *Majungatholus atopus*". *Nature* **422** (6931): 515–518. doi:10.1038/nature01532. PMID 12673249.

[77] Schmitz, L.; Motani, R. (2011). "Nocturnality in Dinosaurs Inferred from Scleral Ring and Orbit Morphology". *Science* **332** (6030): 705–708. Bibcode 2011Sci...332..705S. doi:10.1126/science.1200043. PMID 21493820.

[78] Varricchio DJ, Martin, AJ and Katsura, Y (2007). "First trace and body fossil evidence of a burrowing, denning dinosaur". *Proceedings of the Royal Society B: Biological Sciences* **274** (1616): 1361–1368. doi:10.1098/rspb.2006.0443. PMC 2176205. PMID 17374596.

[79] Chatterjee, S.; Templin, R. J. (2007). "Biplane wing planform and flight performance of the feathered dinosaur *Microraptor gui*" (http://www.pnas.org/cgi/reprint/0609975104v1.pdf) (PDF). *Proceedings of the National Academy of Sciences* **104** (5): 1576–1580. Bibcode 2007PNAS..104.1576C. doi:10.1073/pnas.0609975104. PMC 1780066. PMID 17242354. .

[80] Zhang, F.; Zhou, Z.; Xu, X.; and Wang, X. (2002). "A juvenile coelurosaurian theropod from China indicates arboreal habits". *Naturwissenschaften* **89** (9): 394–398. Bibcode 2002NW.....89..394Z. doi:10.1007/s00114-002-0353-8. PMID 12435090.

[81] Alexander RM (2006). "Dinosaur biomechanics". *Proceedings of the Royal Society of Biological Sciences* **273** (1596): 1849–1855. doi:10.1098/rspb.2006.3532. PMC 1634776. PMID 16822743.

[82] Goriely A & McMillen T (2002). "Shape of a cracking whip". *Physical Review Letters* **88** (24): 244301. Bibcode 2002PhRvL..88x4301G. doi:10.1103/PhysRevLett.88.244301. PMID 12059302.

[83] Henderson, D.M. (2003). "Effects of stomach stones on the buoyancy and equilibrium of a floating crocodilian: A computational analysis". *Canadian Journal of Zoology* **81** (8): 1346–1357. doi:10.1139/z03-122.

[84] Senter, P. (2008). "Voices of the past: a review of Paleozoic and Mesozoic animal sounds". *Historical Biology* **20** (4): 255–287. doi:10.1080/08912960903033327.

[85] Hopson, James A. (1975). "The evolution of cranial display structures in hadrosaurian dinosaurs". *Paleobiology* **1** (1): 21–43.

[86] Diegert, Carl F. (1998). "A digital acoustic model of the lambeosaurine hadrosaur *Parasaurolophus tubicen*". *Journal of Vertebrate Paleontology* **18** (3, Suppl.): 38A.

[87] Parsons, Keith M. (2001). *Drawing out Leviathan: Dinosaurs and the science wars*. Bloomington: Indiana University Press. pp. 22–48. ISBN 0-253-33937-5.

[88] Fisher, P. E., Russell, D. A., Stoskopf, M. K., Barrick, R. E., Hammer, M. & Kuzmitz, A. A. (2000). "Cardiovascular evidence for an intermediate or higher metabolic rate in an ornithischian dinosaur". *Science* **288** (5465): 503–505. Bibcode 2000Sci...288..503F. doi:10.1126/science.288.5465.503. PMID 10775107.

[89] Hillenius, W. J. & Ruben, J. A. (2004). "The evolution of endothermy in terrestrial vertebrates: Who? when? why?". *Physiological and Biochemical Zoology* **77** (6): 1019–1042. doi:10.1086/425185. PMID 15674773.

[90] Rowe T, McBride EF, & Sereno PC (2001). "Dinosaur with a Heart of Stone". *Science* **291** (5505): 783. doi:10.1126/science.291.5505.783a. PMID 11157158.

[91] Chinsamy A, Hillenius WJ (2004). "Physiology of nonavian dinosaurs". *The Dinosauria (2d edition*. University of California Press. pp. 643–659. ISBN 0-520-24209-2.

[92] Kaye, T. G.; Gaugler, G; Sawlowicz, Z; Stepanova, Anna (July 2008). "Dinosaurian Soft Tissues Interpreted as Bacterial Biofilms." (http://www.plosone.org/article/info:doi/10.1371/journal.pone.0002808). *PLoS ONE* **3** (7): e2808. Bibcode 2008PLoSO...3.2808K. doi:10.1371/journal.pone.0002808. PMC 2483347. PMID 18665236. . Retrieved 2008-10-27.

[93] "New Research Challenges Notion That Dinosaur Soft Tissues Still Survive" (http://newswise.com/articles/view/542898/). . Retrieved 2011-05-05.

[94] Wang, H., Yan, Z. and Jin, D. (1 May 1997). "Reanalysis of published DNA sequence amplified from Cretaceous dinosaur egg fossil" (http://mbe.oupjournals.org/cgi/reprint/14/5/589). *Molecular Biology and Evolution* **14** (5): 589–591. PMID 9159936. . Retrieved 2007-12-05.

[95] Chang BS, Jönsson K, Kazmi MA, Donoghue MJ, Sakmar TP (1 September 2002). "Recreating a Functional Ancestral Archosaur Visual Pigment" (http://mbe.oxfordjournals.org/cgi/content/full/19/9/1483). *Molecular Biology and Evolution* **19** (9): 1483–1489.

PMID 12200476. . Retrieved 2007-12-05.

[96] Schweitzer MH, Marshall M, Carron K, Bohle DS, Busse SC, Arnold EV, Barnard D, Horner JR, Starkey JR (1997). "Heme compounds in dinosaur trabecular bone". *Proc Natl Acad Sci U S A.* **94** (12): 6291–6. Bibcode 1997PNAS...94.6291S. doi:10.1073/pnas.94.12.6291. PMC 21042. PMID 9177210.

[97] Embery G, Milner AC, Waddington RJ, Hall RC, Langley MS, Milan AM (2003). "Identification of proteinaceous material in the bone of the dinosaur Iguanodon". *Connect Tissue Res* **44** (Suppl 1): 41–6. doi:10.1080/713713598. PMID 12952172.

[98] Huxley, Thomas H. (1868). "On the animals which are most nearly intermediate between birds and reptiles". *Annals of the Magazine of Natural History* **4** (2): 66–75.

[99] Heilmann, Gerhard (1926). *The Origin of Birds*. London: Witherby. pp. 208pp. ISBN 0-486-22784-7.

[100] Osborn, Henry Fairfield (1924). "Three new Theropoda, *Protoceratops* zone, central Mongolia" (http://digitallibrary.amnh.org/dspace/ bitstream/2246/3223/1/N0144.pdf) (PDF). *American Museum Novitates* **144**: 1–12. .

[101] Ostrom, John H. (1973). "The ancestry of birds". *Nature* **242** (5393): 136. doi:10.1038/242136a0.

[102] Gauthier, Jacques. (1986). "Saurischian monophyly and the origin of birds". In Padian, Kevin. (ed.). *The Origin of Birds and the Evolution of Flight*. Memoirs of the California Academy of Sciences **8**. pp. 1–55.

[103] Mayr, G., Pohl, B. and Peters, D.S. (2005). "A Well-Preserved *Archaeopteryx* Specimen with Theropod Features". *Science* **310** (5753): 1483–1486. Bibcode 2005Sci...310.1483M. doi:10.1126/science.1120331. PMID 16322455.

[104] Martin, Larry D. (2006). "A basal archosaurian origin for birds". *Acta Zoologica Sinica* **50** (6): 977–990.

[105] Feduccia, A. (2002). "Birds are dinosaurs: simple answer to a complex problem". *The Auk* **119**: 1187–1201. doi:10.1642/0004-8038(2002)119[1187:BADSAT]2.0.CO;2.

[106] Wellnhofer, P (1988). "Ein neuer Exemplar von *Archaeopteryx*". *Archaeopteryx* **6**: 1–30.

[107] Xu X.; Norell, M.A.; Kuang X.; Wang X.; Zhao Q.; and Jia C. (2004). "Basal tyrannosauroids from China and evidence for protofeathers in tyrannosauroids". *Nature* **431** (7009): 680–684. Bibcode 2004Natur.431..680X. doi:10.1038/nature02855. PMID 15470426.

[108] Göhlich, U.B.; Chiappe, LM (2006). "A new carnivorous dinosaur from the Late Jurassic Solnhofen archipelago". *Nature* **440** (7082): 329–332. Bibcode 2006Natur.440..329G. doi:10.1038/nature04579. PMID 16541071.

[109] Lingham-Soliar, T. (2003). "The dinosaurian origin of feathers: perspectives from dolphin (Cetacea) collagen fibers". *Naturwissenschaften* **12** (12): 563–567. Bibcode 2003NW.....90..563L. doi:10.1007/s00114-003-0483-7. PMID 14676953.

[110] Feduccia, A.; Lingham-Soliar, T; Hinchliffe, JR (2005). "Do feathered dinosaurs exist? Testing the hypothesis on neontological and paleontological evidence". *Journal of Morphology* **266** (2): 125–166. doi:10.1002/jmor.10382. PMID 16217748.

[111] Lingham-Soliar, T.; Feduccia, A; Wang, X (2007). "A new Chinese specimen indicates that 'protofeathers' in the Early Cretaceous theropod dinosaur *Sinosauropteryx* are degraded collagen fibres". *Proceedings of the Biological Sciences* **274** (1620): 1823–9. doi:10.1098/rspb.2007.0352. PMC 2270928. PMID 17521978.

[112] Prum, Richard O. (April 2003). "Are Current Critiques Of The Theropod Origin Of Birds Science? Rebuttal To Feduccia 2002". *The Auk* **120** (2): 550–61. doi:10.1642/0004-8038(2003)120[0550:ACCOTT]2.0.CO;2. JSTOR 4090212.

[113] O'Connor, P.M. & Claessens, L.P.A.M. (2005). "Basic avian pulmonary design and flow-through ventilation in non-avian theropod dinosaurs". *Nature* **436** (7048): 253–256. Bibcode 2005Natur.436..253O. doi:10.1038/nature03716. PMID 16015329.

[114] Sereno, P.C.; Martinez, RN; Wilson, JA; Varricchio, DJ; Alcober, OA; Larsson, HC; Kemp, Tom (September 2008). "Evidence for Avian Intrathoracic Air Sacs in a New Predatory Dinosaur from Argentina" (http://www.plosone.org/article/info:doi/10.1371/journal.pone. 0003303). *PLoS ONE* **3** (9): e3303. Bibcode 2008PLoSO...3.3303S. doi:10.1371/journal.pone.0003303. PMC 2553519. PMID 18825273. . Retrieved 2008-10-27.

[115] "Meat-Eating Dinosaur from Argentina Had Bird-Like Breathing System" (http://newswise.com/articles/view/544722/). . Retrieved 2011-05-05.

[116] Schweitzer, M.H.; Wittmeyer, JL; Horner, JR (2005). "Gender-specific reproductive tissue in ratites and *Tyrannosaurus rex*". *Science* **308** (5727): 1456–1460. Bibcode 2005Sci...308.1456S. doi:10.1126/science.1112158. PMID 15933198.

[117] Lee, Andrew H.; Werning, S (2008). "Sexual maturity in growing dinosaurs does not fit reptilian growth models" (http://www.pnas.org/ cgi/content/abstract/105/2/582). *Proceedings of the National Academy of Sciences* **105** (2): 582–587. Bibcode 2008PNAS..105..582L. doi:10.1073/pnas.0708903105. PMC 2206579. PMID 18195356. .

[118] Xu, X. and Norell, M.A. (2004). "A new troodontid dinosaur from China with avian-like sleeping posture". *Nature* **431** (7010): 838–841. Bibcode 2004Natur.431..838X. doi:10.1038/nature02898. PMID 15483610.

[119] Norell M.A., Clark J.M., Chiappe L.M., Dashzeveg D. (1995). "A nesting dinosaur". *Nature* **378** (6559): 774–776. Bibcode 1995Natur.378..774N. doi:10.1038/378774a0.

[120] Varricchio, D. J.; Moore, J. R.; Erickson, G. M.; Norell, M. A.; Jackson, F. D.; Borkowski, J. J. (2008). "Avian Paternal Care Had Dinosaur Origin". *Science* **322**: 1826. Bibcode 2008Sci...322.1826V. doi:10.1126/science.1163245.

[121] Wings O (2007). "A review of gastrolith function with implications for fossil vertebrates and a revised classification" (http://www.app. pan.pl/archive/published/app52/app52-001.pdf) (PDF). *Palaeontologica Polonica* **52** (1): 1–16. . Retrieved 2011-05-05.

[122] Dingus, L. and Rowe, T. (1998). *The Mistaken Extinction – Dinosaur Evolution and the Origin of Birds*. New York: W. H. Freeman.

[123] Miller KG, Kominz MA, Browning JV, Wright JD, Mountain GS, Katz ME, Sugarman PJ, Cramer BS, Christie-Blick N, Pekar SF (2005). "The Phanerozoic record of global sea-level change". *Science* **310** (5752): 1293–8. Bibcode 2005Sci...310.1293M. doi:10.1126/science.1116412. PMID 16311326.

[124] McArthura JM, Janssenb NMM, Rebouletc S, Lengd MJ, Thirlwalle MF & van de Shootbruggef B (2007). "Palaeotemperatures, polar ice-volume, and isotope stratigraphy (Mg/Ca, δ18O, δ13C, 87Sr/86Sr): The Early Cretaceous (Berriasian, Valanginian, Hauterivian)". *Palaeogeography, Palaeoclimatology, Palaeoecology* **248** (3–4): 391–430. doi:10.1016/j.palaeo.2006.12.015.

[125] Alvarez, LW, Alvarez, W, Asaro, F, and Michel, HV (1980). "Extraterrestrial cause for the Cretaceous–Tertiary extinction". *Science* **208** (4448): 1095–1108. Bibcode 1980Sci...208.1095A. doi:10.1126/science.208.4448.1095. PMID 17783054.

[126] Hildebrand, Alan R.; Penfield, Glen T.; Kring, David A.; Pilkington, Mark; Zanoguera, Antonio Camargo; Jacobsen, Stein B.; Boynton, William V. (September 1991). "Chicxulub Crater; a possible Cretaceous/Tertiary boundary impact crater on the Yucatan Peninsula, Mexico". *Geology* **19** (9): 867–871. Bibcode 1991Geo....19..867H. doi:10.1130/0091-7613(1991)019<0867:CCAPCT>2.3.CO;2.

[127] Pope KO, Ocampo AC, Kinsland GL, Smith R (1996). "Surface expression of the Chicxulub crater". *Geology* **24** (6): 527–30. Bibcode 1996Geo....24..527P. doi:10.1130/0091-7613(1996)024<0527:SEOTCC>2.3.CO;2. PMID 11539331.

[128] Robertson, D.S.; et al. (30 September). "Survival in the first hours of the Cenozoic" (http://webh01.ua.ac.be/funmorph/raoul/ macroevolutie/Robertson2004.pdf). *Geological Society of America Bulletin* **116** (5/6): 760–768. doi:10.1130/B25402.1. . Retrieved 15 June 2011.

[129] P, Claeys; Goderis, S (2007-09-05). "Solar System: Lethal billiards". *Nature* **449** (7158): 30–31. Bibcode 2007Natur.449...30C. doi:10.1038/449030a. PMID 17805281.

[130] Plotner, Tammy (2011). "Did Asteroid Baptistina Kill the Dinosaurs? Think other WISE..." (http://www.universetoday.com/89050/ did-asteroid-baptistina-kill-the-dinosaurs-think-other-wise/#more-89050). *Universe Today.* . Retrieved 2011-9-20.

[131] edited by Christian Koeberl and Kenneth G. MacLeod. (2002). *Catastrophic Events and Mass Extinctions.* Geological Society of America. ISBN 0-8137-2356-6. OCLC 213836505.

[132] Hofman, C, Féraud, G & Courtillot, V (2000). "40Ar/39Ar dating of mineral separates and whole rocks from the Western Ghats lava pile: further constraints on duration and age of the Deccan traps". *Earth and Planetary Science Letters* **180**: 13–27. Bibcode 2000E&PSL.180...13H. doi:10.1016/S0012-821X(00)00159-X.

[133] Duncan, RA & Pyle, DG (1988). "Rapid eruption of the Deccan flood basalts at the Cretaceous/Tertiary boundary". *Nature* **333** (6176): 841–843. Bibcode 1988Natur.333..841D. doi:10.1038/333841a0.

[134] Alvarez, W (1997). *T. rex and the Crater of Doom.* Princeton University Press. pp. 130–146. ISBN 978-0-691-01630-6.

[135] Lloyd, G.T., Davis, K.E., Pisani, D. (22 July 2008). "Dinosaurs and the Cretaceous Terrestrial Revolution" (http://journals.royalsociety. org/content/7k63203q852h4006/). *Proceedings of the Royal Society: Biology* **275** (1650): 2483–90. doi:10.1098/rspb.2008.0715. PMC 2603200. PMID 18647715. . Retrieved 2008-07-28.

[136] Fassett, JE, Lucas, SG, Zielinski, RA, and Budahn, JR (2001). "Compelling new evidence for Paleocene dinosaurs in the Ojo Alamo Sandstone, San Juan Basin, New Mexico and Colorado, USA" (http://www.lpi.usra.edu/meetings/impact2000/pdf/3139.pdf) (PDF). *Catastrophic events and mass extinctions, Lunar and Planetary Contribution* **1053**: 45–46. . Retrieved 2007-05-18.

[137] Sloan, R. E., Rigby, K,. Van Valen, L. M., Gabriel, Diane (1986). "Gradual dinosaur extinction and simultaneous ungulate radiation in the Hell Creek Formation". *Science* **232** (4750): 629–633. Bibcode 1986Sci...232..629S. doi:10.1126/science.232.4750.629. PMID 17781415.

[138] Fastovsky, David E.; Sheehan, Peter M. (2005). "Reply to comment on "The Extinction of the dinosaurs in North America"" (http://www. geosociety.org/gsatoday/archive/15/7/pdf/i1052-5173-15-7-11b.pdf) (PDF). *GSA Today* **15**: 11. doi:10.1130/1052-5173(2005)015[11b:RTEOTD]2.0.CO;2. .

[139] Sullivan, RM (2003). "No Paleocene dinosaurs in the San Juan Basin, New Mexico" (http://gsa.confex.com/gsa/2003RM/ finalprogram/abstract_47695.htm). *Geological Society of America Abstracts with Programs* **35** (5): 15. . Retrieved 2007-07-02.

[140] Fassett J.E., Heaman L.M., Simonetti A. (2011). "Direct U–Pb dating of Cretaceous and Paleocene dinosaur bones, San Juan Basin, New Mexico". *Geology* **39**: 159–162. doi:10.1130/G31466.1.

[141] Dong Zhiming (1992). *Dinosaurian Faunas of China.* China Ocean Press, Beijing. ISBN 3-540-52084-8. OCLC 26522845.

[142] "Dinosaur bones 'used as medicine'" (http://news.bbc.co.uk/2/hi/asia-pacific/6276948.stm). BBC News. 2007-07-06. . Retrieved 2007-07-06.

[143] Sarjeant WAS (1997). "The earliert discoveries". In Farlow JO, Brett-Surman MK. *The Complete Dinosaur.* Bloomington: Indiana University Press. pp. 3–11. ISBN 0-253-33349-0.

[144] Lhuyd, E. (1699). *Lithophylacii Britannici Ichnographia, sive lapidium aliorumque fossilium Britannicorum singulari figura insignium.* Gleditsch and Weidmann:London.

[145] Delair, J.B.; Sarjeant, W.A.S. (2002). "The earliest discoveries of dinosaurs: the records re-examined". *Proceedings of the Geologists' Association* **113**: 185–197.

[146] Gunther RT (1968). *Life and letters of Edward Lhwyd,: Second keeper of the Museum Ashmoleanum (Early science in Oxford Volume XIV).* Dawsons of Pall Mall.

[147] Buckland W (1824). "Notice on the *Megalosaurus* or great Fossil Lizard of Stonesfield.". *Transactions of the Geological Society of London* **1**: 390–396.

[148] http://collections.tepapa.govt.nz/objectdetails.aspx?oid=212194&coltype=history®no=gh004839

[149] Mantell, Gideon A. (1825). "Notice on the Iguanodon, a newly discovered fossil reptile, from the sandstone of Tilgate forest, in Sussex.". *Philosophical Transactions of the Royal Society* **115**: 179–186. doi:10.1098/rstl.1825.0010. JSTOR 107739.

[150] Sues, Hans-Dieter (1997). "European Dinosaur Hunters". In Farlow JO, Brett-Surman MK. *The Complete Dinosaur.* Bloomington: Indiana University Press. p. 14. ISBN 0-253-33349-0.

[151] Holmes T (1996). *Fossil Feud: The Bone Wars of Cope and Marsh, Pioneers in Dinosaur Science*. Silver Burdett Press. ISBN 978-0-382-39147-7. OCLC 34472600.

[152] Torrens, H.S. (1993). "The dinosaurs and dinomania over 150 years". *Modern Geology* **18** (2): 257–286.

[153] Breithaupt, Brent H. (1997). "First golden period in the USA". In Currie, Philip J. and Padian, Kevin (eds.). *The Encyclopedia of Dinosaurs*. San Diego: Academic Press. pp. 347–350. ISBN 978-0-12-226810-6.

[154] *"London. Michaelmas term lately over, and the Lord Chancellor sitting in Lincoln's Inn Hall. Implacable November weather. As much mud in the streets, as if the waters had but newly retired from the face of the earth, and it would not be wonderful to meet a Megalosaurus, forty feet long or so, waddling like an elephantine lizard up Holborne Hill."* Dickens CJH (1852). *Bleak House*. London: Bradbury & Evans. p. 1.

[155] Glut, D.F., and Brett-Surman, M.K. (1997). Farlow, James O. and Brett-Surman, Michael K. (eds.). ed. *The Complete Dinosaur*. Indiana University Press. pp. 675–697. ISBN 978-0-253-21313-6.

Further reading

- Bakker, Robert T. (1986). *The Dinosaur Heresies: New Theories Unlocking the Mystery of the Dinosaurs and Their Extinction*. New York: Morrow. ISBN 0-688-04287-2.
- Holtz, Thomas R. Jr. (2007). *Dinosaurs: The Most Complete, Up-to-Date Encyclopedia for Dinosaur Lovers of All Ages*. New York: Random House. ISBN 978-0-375-82419-7.
- Paul, Gregory S. (2000). *The Scientific American Book of Dinosaurs*. New York: St. Martin's Press. ISBN 0-312-26226-4.
- Paul, Gregory S. (2002). *Dinosaurs of the Air: The Evolution and Loss of Flight in Dinosaurs and Birds*. Baltimore: The Johns Hopkins University Press. ISBN 0-8018-6763-0.

External links

General

- DinoDatabase.com | Hundreds of dinosaurs and dinosaur related topics. (http://www.dinodatabase.com/)

Images

- The Science and Art of Gregory S. Paul (http://www.gspauldino.com/) Influential paleontologist's anatomy art and paintings
- Skeletal Drawing (http://skeletaldrawing.com/) Professional restorations of numerous dinosaurs, and discussions of dinosaur anatomy.

Video

- BBC Nature: Watch dinosaurs brought to life and get experts' interpretations with videos from BBC programmes including Walking with Dinosaurs (http://www.bbc.co.uk/nature/life/dinosaur)

Popular

- Dinosaurs & other extinct creatures (http://www.nhm.ac.uk/nature-online/life/dinosaurs-other-extinct-creatures/index.html): From the Natural History Museum, a well illustrated dinosaur directory.
- Dinosaurnews (http://www.dinosaurnews.org/) (*www.dinosaurnews.org*) The dinosaur-related headlines from around the world. Recent news on dinosaurs, including finds and discoveries, and many links.
- Dinosauria (http://www.ucmp.berkeley.edu/diapsids/dinosaur.html) From UC Berkeley Museum of Paleontology Detailed information – scroll down for menu.
- LiveScience.com (http://www.livescience.com/dinosaurs/) All about dinosaurs, with current featured articles.
- Zoom Dinosaurs (http://www.enchantedlearning.com/subjects/dinosaurs/) (*www.enchantedlearning.com*) From Enchanted Learning. Kids' site, info pages and stats, theories, history.
- Dinosaur genus list (http://www.geol.umd.edu/~tholtz/dinoappendix/HoltzappendixWinter2010.pdf) contains data tables on nearly every published Mesozoic dinosaur genus as of January 2011.
- LiveScience.com (http://www.livescience.com/animals/090621-dinosaur-size.html) Giant Dinosaurs Get Downsized by *LiveScience*, June 21, 2009

Technical

- *Palaeontologia Electronica* (http://palaeo-electronica.org/) From Coquina Press. Online technical journal.
- Dinobase (http://dinobase.gly.bris.ac.uk/) A searchable dinosaur database, from the University of Bristol, with dinosaur lists, classification, pictures, and more.
- DinoData (http://www.dinodata.org/index.php) (*www.dinodata.org*) Technical site, essays, classification, anatomy.
- Dinosauria On-Line (http://www.dinosauria.com/dml/dml.htm) (*www.dinosauria.com*) Technical site, essays, pronunciation, dictionary.
- Thescelosaurus! (http://www.thescelosaurus.com/) By Justin Tweet. Includes a cladogram and small essays on each relevant genera and species.
- Dinosauromorpha Cladogram (http://www.palaeos.com/Vertebrates/Units/Unit310/000.html) From Palaeos (http://www.Palaeos.com/). A detailed amateur site about all things paleo.

Genus

In biology, a **genus** (plural: **genera**) is a low-level taxonomic rank used in the biological classification of living and fossil organisms, which is an example of definition by genus and differentia. Genera and higher taxonomic levels such as families are used in biodiversity studies, particularly in fossil studies since species cannot always be confidently identified and genera and families typiclaly have longer stratigraphic ranges than species.[1]

The term comes from Latin genus "descent, family, type, gender",[2] cognate with Greek: γένος – *genos*, "race, stock, kin".[3]

The composition of a genus is determined by a taxonomist. The standards for genus classification are not strictly codified, and hence different authorities often produce different classifications for genera. In the hierarchy of the binomial classification system, genus comes above species and below family.

Generic name

The scientific name of a genus may be called the **generic name** or **generic epithet**: it is always capitalized. It plays a pivotal role in binomial nomenclature, the system of biological nomenclature.

Binomial nomenclature

The rules for scientific names are laid down in the Nomenclature Codes; depending on the kind of organism and the Kingdom it belongs to, a different Code may apply, with different rules, laid down in a different terminology. The advantages of scientific over common names are that they are accepted by speakers of all languages, and that each species has only one name. This reduces the confusion that may arise from the use of a common name to designate different things in different places (example elk), or from the existence of several common names for a single species.

It is possible for a genus to be assigned to a kingdom governed by one particular Nomenclature Code by one taxonomist, while other taxonomists assign it to a kingdom governed by a different Code, but this is the exception, not the rule.

Pivotal in binomial nomenclature

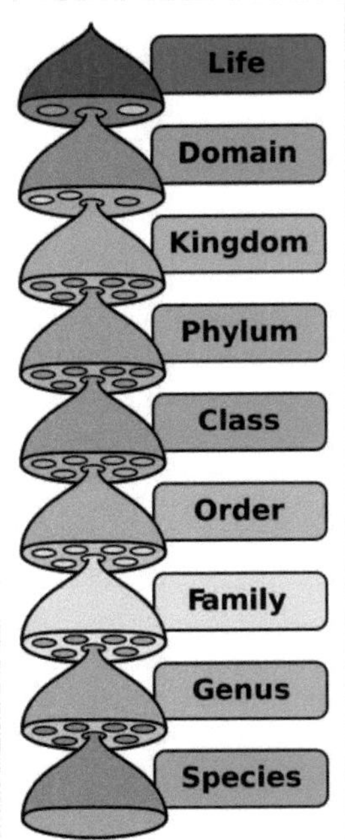

The hierarchy of biological classification's eight major taxonomic ranks, which is an example of definition by genus and differentia. A family contains one or more genera. Intermediate minor rankings are not shown.

The generic name often is a component of the names of taxa of lower rank. For example, *Canis lupus* is the scientific name of the Gray wolf, a species, with *Canis* the generic name for the dog and its close relatives, and with *lupus* particular (specific) for the wolf (*lupus* is written in lower case). Similarly, *Canis lupus familiaris* is the scientific name for the domestic dog.

Taxonomic units in higher ranks often have a name that is based on a generic name, such as the family name Canidae, which is based on *Canis*. However, not all names in higher ranks are necessarily based on the name of a genus: for example, Carnivora is the name for the order to which the dog belongs.

The problem of identical names used for different genera

A genus in one kingdom is allowed to bear a scientific name that is in use as a generic name (or the name of a taxon in another rank) in a kingdom that is governed by a different nomenclature code. Although this is discouraged by both the International Code of Zoological Nomenclature and the International Code of Botanical Nomenclature, there are some five thousand such names that are in use in more than one kingdom. For instance, *Anura* is the name of the order of frogs but also is the name of a genus of plants (although not current: it is a synonym); *Aotus* is the genus of golden peas and night monkeys; *Oenanthe* is the genus of wheatears and water dropworts, *Prunella* is the genus of accentors and self-heal, and *Proboscidea* is the order of elephants and the genus of devil's claws.

Within the same kingdom one generic name can apply to only one genus. This explains why the platypus genus is named *Ornithorhynchus*—George Shaw named it *Platypus* in 1799, but the name *Platypus* had already been given to a group of ambrosia beetles by Johann Friedrich Wilhelm Herbst in 1793. Names with the same form but applying to different taxa are called homonyms. Since beetles and platypuses are both members of the kingdom Animalia, the name *Platypus* could not be used for both. Johann Friedrich Blumenbach published the replacement name *Ornithorhynchus* in 1800.

Types and genera

Because of the rules of scientific naming, or "binomial nomenclature", each genus should have a designated type, although in practice there is a backlog of older names that may not yet have a type. In zoology this is the type species; the generic name is permanently associated with the type specimen of its type species. Should this specimen turn out to be assignable to another genus, the generic name linked to it becomes a junior synonym, and the remaining taxa in the former genus need to be reassessed.

See scientific classification and nomenclature codes for more details of this system. Also see type genus.

Guidelines

There are no hard and fast rules that a taxonomist has to follow in deciding what does and what does not belong in a particular genus. This does not mean that there is no common ground among taxonomists in what constitutes a "good" genus. For instance, some rules-of-thumb for delimiting a genus are outlined in Gill.[4] According to these, a genus should fulfill three criteria to be descriptively useful:

1. monophyly – all descendants of an ancestral taxon are grouped together;
2. reasonable compactness – a genus should not be expanded needlessly; and
3. distinctness – in regards of evolutionarily relevant criteria, i.e. ecology, morphology, or biogeography; note that DNA sequences are a *consequence* rather than a *condition* of diverging evolutionary lineages except in cases where they directly inhibit gene flow (e.g. postzygotic barriers).

Nomenclature

...difficulties occurring in generic nomenclature: similar cases abound, and become complicated by the different views taken of the matter by the various taxonomists.

Prof. C. S. Rafinesque. 1836[5]

None of the nomenclature codes require such criteria for defining a genus, because these are concerned with the nomenclature rules, not with taxonomy. These regulate formal nomenclature, aiming for universal and stable scientific names.

See also

- Andrew Delmar Hopkins
- List of the largest genera of flowering plants

References

[1] Sahney, S., Benton, M.J. and Ferry, P.A. (2010). "Links between global taxonomic diversity, ecological diversity and the expansion of vertebrates on land" (http://rsbl.royalsocietypublishing.org/content/6/4/544.full.pdf+html) (PDF). *Biology Letters* **6** (4): 544–547. doi:10.1098/rsbl.2009.1024. PMC 2936204. PMID 20106856. .

[2] Merriam Webster Dictionary (http://www.merriam-webster.com/dictionary/genus)

[3] Genos (http://www.perseus.tufts.edu/cgi-bin/ptext?doc=Perseus:text:1999.04.0057:entry=#21921), Henry George Liddell, Robert Scott, 'A Greek-English Lexicon, *at Perseus*

[4] Gill, F. B., B. Slikas, and F. H. Sheldon. "Phylogeny of titmice (Paridae): II. Species relationships based on sequences of the mitochondrial cytochrome-b gene." Auk 122(1): 121-143, 2005. (Google Scholar) (http://scholar.google.com/scholar?cluster=16219444564703958615&hl=en)

[5] Rafinesque, Prof. C. S. (1836). "Generic Rules" (http://www.us.archive.org/GnuBook/?id=floratelluriana00rafi#99). *Flora telluriana Pars Prima First Part of the Synoptical Flora Telluriana, Centuries I, II, III, IV. With new Natural Classes, Orders and families: containing*

the 2000 New or revised Genera and Species of Trees, Palms, Shrubs, Vines, Plants, Lilies, Grasses, Ferns, Algas, Fungi, & c. from North and South America, Polynesia, Australia, Asia Europe and Africa, omitted or mistaken by the authors, that were observed or ascertained, described or revised, collected or figured, between 1796 and 1836. (http://www.us.archive.org/GnuBook/?id=floratelluriana00rafi#13). **1.** Philadelphia: H. Probasco. . Retrieved 2009-04-02. "...difficulties occurring in generic nomenclature: similar cases abound, and become complicated by the different views taken of the matter by the various botanists."

External links

- Nomenclator Zoologicus (http://uio.mbl.edu/NomenclatorZoologicus/): Index of all genus and subgenus names in zoological nomenclature from 1758 to 2004.
- Fauna Europaea Database for Taxonomy (http://www.faunaeur.org/full_results.php?id=193482)

rue:Рід (біологія)

Geologic time scale

The **geologic time scale** provides a system of chronologic measurement relating stratigraphy to time that is used by geologists, paleontologists and other earth scientists to describe the timing and relationships between events that have occurred during the history of the Earth. The table of geologic time spans presented here agrees with the dates and nomenclature proposed by the International Commission on Stratigraphy, and uses the standard color codes of the United States Geological Survey.

Evidence from radiometric dating indicates that the Earth is about 4.570 billion years old. The geological or *deep time* of Earth's past has been organized into various units according to events which took place in each period. Different spans of time on the time scale are usually delimited by major geological or paleontological events, such as mass extinctions. For example, the boundary between the Cretaceous period and the Paleogene period is defined by the Cretaceous–Tertiary extinction event, which marked the demise of the dinosaurs and of many marine species. Older periods which predate the reliable fossil record are defined by absolute age.

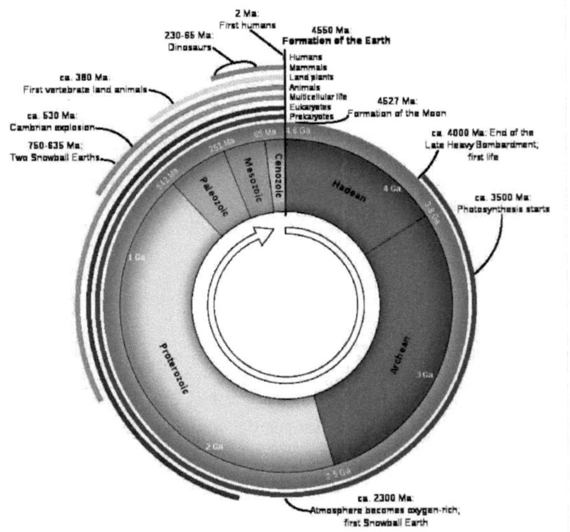

This clock representation shows some of the major units of geological time and definitive events of Earth history. The Hadean eon represents the time before fossil record of life on Earth; its upper boundary is now regarded as 4.0 Ga.[1] Other subdivisions reflect the evolution of life; the Archean and Proterozoic are both eons, the Palaeozoic, Mesozoic and Cenozoic are eras of the Phanerozoic eon. The two million year Quaternary period, the time of recognizable humans, is too small to be visible at this scale.

Each era on the scale is separated from the next by a major event or change.

Terminology

e [2] h [2]		
Units in geochronology and stratigraphy[3]		
Segments of rock (strata) in chronostratigraphy	**Periods of time in geochronology**	**Notes**
Eonothem	Eon	4 total, half a billion years or more
Erathem	Era	12 total, several hundred million years
System	Period	
Series	Epoch	tens of millions of years
Stage	Age	millions of years
Chronozone	Chron	smaller than an age/stage, not used by the ICS timescale

The largest defined unit of time is the **supereon**, composed of **eons**. Eons are divided into **eras**, which are in turn divided into **periods, epochs** and **ages**. The terms eonothem, erathem, system, series, and stage are used to refer to the layers of rock that correspond to these periods of geologic time.

Geologists qualify these units as Early, Mid, and Late when referring to time, and Lower, Middle, and Upper when referring to the corresponding rocks. For example, the Lower Jurassic Series in chronostratigraphy corresponds to the Early Jurassic Epoch in geochronology.[4] The adjectives are capitalized when the subdivision is formally recognized, and lower case when not; thus "early Miocene" but "Early Jurassic."

Geologic units from the same time but different parts of the world often look different and contain different fossils, so the same period was historically given different names in different locales. For example, in North America the Lower Cambrian is called the Waucoban series that is then subdivided into zones based on succession of trilobites. In East Asia and Siberia, the same unit is split into Alexian, Atdabanian, and Botomian stages. A key aspect of the work of the International Commission on Stratigraphy is to reconcile this conflicting terminology and define universal horizons that can be used around the world.[5]

Graphical timelines

The second and third timelines are each subsections of their preceding timeline as indicated by asterisks.

Millions of Years

The Holocene (the latest epoch) is too short to be shown clearly on this timeline.

History of the time scale and names

In classical antiquity, Aristotle saw that fossil seashells from rocks were similar to those found on the beach and inferred that the fossils were once part of living animals. He reasoned that the positions of land and sea had changed over long periods of time. Leonardo da Vinci concurred with Aristotle's view that fossils were the remains of ancient life.[6]

The 11th-century Persian geologist Avicenna (Ibn Sina) examined various fossils and inferred that they originated from the

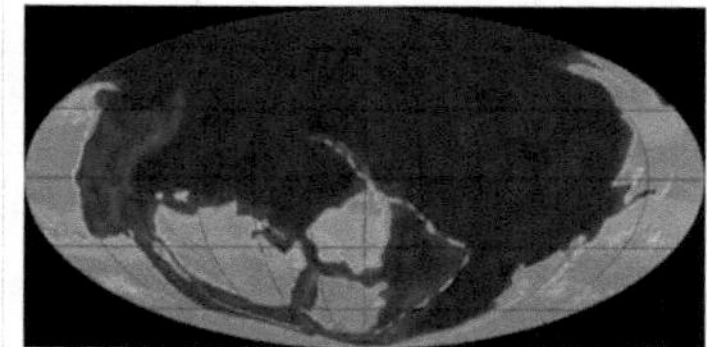
Animation showing Earth's palaeogeographic reconstruction beginning from early Cambrian period.

petrifaction of plants and animals.[7] He also first proposed one of the principles underlying geologic time scales: the law of superposition of strata. While discussing the origins of mountains in *The Book of Healing* in 1027, he outlined the principle as follows:

Diagram of geological time scale, where the past is toward the bottom of the spiral

> It is also possible that the sea may have happened to flow little by little over the land consisting of both plain and mountain, and then have ebbed away from it. ... It is possible that each time the land was exposed by the ebbing of the sea a layer was left, since we see that some mountains appear to have been piled up layer by layer, and it is therefore likely that the clay from which they were formed was itself at one time arranged in layers. One layer was formed first, then at a different period, more layers were formed and piled, upon the first, and so on. Over each layer there spread a substance of different material, which formed a partition between it and the next layer; but when petrification took place something occurred to the partition which caused it to break up and disintegrate from between the layers [possibly referring to unconformity]. ... As to the beginning of the sea, its clay is either sedimentary or primeval, the latter not being sedimentary. It is probable that the sedimentary clay was formed by the disintegration of the strata of mountains. Such is the formation of mountains.[8]

Later in the 11th century, the Chinese naturalist, Shen Kuo (1031–1095), also recognized the concept of 'deep time'.[9]

The principles underlying geologic (geological) time scales were later laid down by Nicholas Steno in the late 17th century. Steno argued that rock layers (or strata) are laid down in succession, and that each represents a "slice" of time. He also formulated the law of superposition, which states that any given stratum is probably older than those above it and younger than those below it. While Steno's principles were simple, applying them to real rocks proved complex. Over the course of the 18th century geologists realized that:

1. Sequences of strata were often eroded, distorted, tilted, or even inverted after deposition;
2. Strata laid down at the same time in different areas could have entirely different appearances;
3. The strata of any given area represented only part of the Earth's long history.

The first serious attempts to formulate a geological time scale that could be applied anywhere on Earth were made in the late 18th century. The most influential of those early attempts (championed by Abraham Werner, among others) divided the rocks of the Earth's crust into four types: Primary, Secondary, Tertiary, and Quaternary. Each type of rock, according to the theory, formed during a specific period in Earth history. It was thus possible to speak of a "Tertiary Period" as well as of "Tertiary Rocks." Indeed, "Tertiary" (now Paleocene - Pliocene) and "Quaternary" (now Pleistocene and Holocene) remained in use as names of geological periods well into the 20th century.

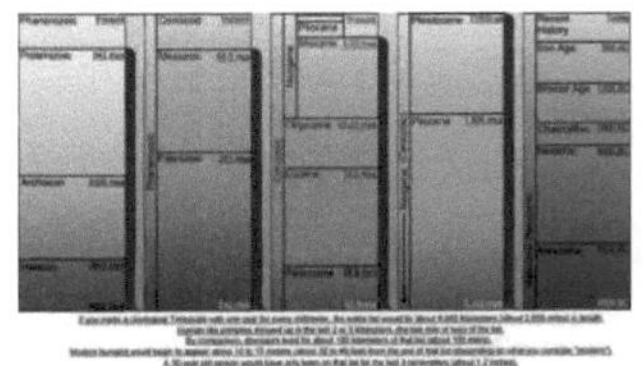

A comparative geological timescale

The Neptunist theories popular at this time (expounded by Werner) proposed that all rocks had precipitated out of a single enormous flood. A major shift in thinking came when James Hutton presented his *Theory of the Earth; or, an Investigation of the Laws Observable in the Composition, Dissolution, and Restoration of Land Upon the Globe* before the Royal Society of Edinburgh in March and April 1785. It has been said that "as things appear from the perspective of the twentieth century, James Hutton in those reading became the founder of modern geology"[10]

Hutton proposed that the interior of the Earth was hot, and that this heat was the engine which drove the creation of new rock: land was eroded by air and water and deposited as layers in the sea; heat then consolidated the sediment into stone, and uplifted it into new lands. This theory was dubbed "Plutonist" in contrast to the flood-oriented theory.

The identification of strata by the fossils they contained, pioneered by William Smith, Georges Cuvier, Jean d'Omalius d'Halloy, and Alexandre Brogniart in the early 19th century, enabled geologists to divide Earth history more precisely. It also enabled them to correlate strata across national (or even continental) boundaries. If two strata (however distant in space or different in composition) contained the same fossils, chances were good that they had been laid down at the same time. Detailed studies between 1820 and 1850 of the strata and fossils of Europe produced the sequence of geological periods still used today.

The process was dominated by British geologists, and the names of the periods reflect that dominance. The "Cambrian", (the Roman name for Wales) and the "Ordovician", and "Silurian", named after ancient Welsh tribes, were periods defined using stratigraphic sequences from Wales.[11] The "Devonian" was named for the English county of Devon, and the name "Carboniferous" was simply an adaptation of "the Coal Measures", the old British geologists' term for the same set of strata. The "Permian" was named after Perm, Russia, because it was defined using strata in that region by Scottish geologist Roderick Murchison. However, some periods were defined by geologists from other countries. The "Triassic" was named in 1834 by a German geologist Friedrich Von Alberti from the three distinct layers (Latin *trias* meaning triad) —red beds, capped by chalk, followed by black shales— that are found throughout Germany and Northwest Europe, called the 'Trias'. The "Jurassic" was named by a French geologist Alexandre Brogniart for the extensive marine limestone exposures of the Jura Mountains. The "Cretaceous" (from Latin *creta* meaning 'chalk') as a separate period was first defined by Belgian geologist Jean d'Omalius d'Halloy in 1822, using strata in the Paris basin[12] and named for the extensive beds of chalk (calcium carbonate deposited by the shells of marine invertebrates).

British geologists were also responsible for the grouping of periods into Eras and the subdivision of the Tertiary and Quaternary periods into epochs.

When William Smith and Sir Charles Lyell first recognized that rock strata represented successive time periods, time scales could be estimated only very imprecisely since various kinds of rates of change used in estimation were highly variable. While creationists had been proposing dates of around six or seven thousand years for the age of the Earth based on the Bible, early geologists were suggesting millions of years for geologic periods with some even suggesting a virtually infinite age for the Earth. Geologists and paleontologists constructed the geologic table based on the relative positions of different strata and fossils, and estimated the time scales based on studying rates of various kinds of weathering, erosion, sedimentation, and lithification. Until the discovery of radioactivity in 1896 and the development of its geological applications through radiometric dating during the first half of the 20th century (pioneered by such geologists as Arthur Holmes) which allowed for more precise absolute dating of rocks, the ages of various rock strata and the age of the Earth were the subject of considerable debate.

The first geologic time scale was eventually published in 1913 by the British geologist Arthur Holmes.[13] He greatly furthered the newly created discipline of geochronology and published the world renowned book *The Age of the Earth* in which he estimated the Earth's age to be at least 1.6 billion years.[14]

In 1977, the *Global Commission on Stratigraphy* (now the International Commission on Stratigraphy) started an effort to define global references (Global Boundary Stratotype Sections and Points) for geologic periods and faunal stages. The commission's most recent work is described in the 2004 geologic time scale of Gradstein et al.[15] A UML model for how the timescale is structured, relating it to the GSSP, is also available.[16]

Table of geologic time

The following table summarizes the major events and characteristics of the periods of time making up the geologic time scale. As above, this time scale is based on the International Commission on Stratigraphy. (See lunar geologic timescale for a discussion of the geologic subdivisions of Earth's moon.) This table is arranged with the most recent geologic periods at the top, and the most ancient at the bottom. The height of each table entry does not correspond to the duration of each subdivision of time.

The content of the table is based on the current official geologic time scale of the International Commission on Stratigraphy,[17] with the epoch names altered to the early/late format from lower/upper as recommended by the ICS when dealing with chronostratigraphy.[4]

See also

- Age of the Earth
- Anthropocene/Homogenocene
- Bubnoff unit
- Deep time
- Evolutionary history of life
- Geological history of Earth
- Geology of Mars/areology
- Graphical timeline of the universe
- History of the Earth
- List of fossil sites *(with link directory)*
- Logarithmic timeline
- Lunar geologic timescale/selenological timescale
- Natural history
- New Zealand geologic time scale
- Prehistoric life
- Timeline of the Big Bang
- Timeline of evolution
- Timeline of the geologic history of the United States
- Timeline of human evolution
- Timeline of the Precambrian

References and footnotes

[1] International Commission on Stratigraphy 2008: stratigraphy.org (http://www.stratigraphy.org/column.php?id=Chart/Time Scale), retrieved 9 March 2009.

[2] http://en.wikipedia.org/wiki/Template%3Ageology_to_paleobiology

[3] International Commission on Stratigraphy. "International Stratigraphic Chart" (http://www.stratigraphy.org/upload/ISChart2009.pdf). . Retrieved 2009-09-25.

[4] International Commission on Stratigraphy. "Chronostratigraphic Units." *International Stratigraphic Guide*. Accessed 14 December 2009. stratigraphy.org (http://www.stratigraphy.org/upload/bak/chron.htm)

[5] Statutes of the International Commission on Stratigraphy (http://www.stratigraphy.org/bak/status.htm#_2. __PURPOSE_AND_OBJECTIVES), retrieved 26 November 2009

[6] Correlating Earth's History (http://www.wmnh.com/wmas0002.htm), Paul R. Janke

[7] Rudwick, M. J. S. (1985). *The Meaning of Fossils: Episodes in the History of Palaeontology*. University of Chicago Press. p. 24. ISBN 0226731030.

[8] Quoted in The contribution of Ibn Sina (Avicenna) to the development of the Earth Sciences (http://www.muslimheritage.com/uploads/ibnsina.pdf), among other sources

[9] Sivin, Nathan (1995). *Science in Ancient China: Researches and Reflections*. Brookfield, Vermont: Ashgate Publishing Variorum series. III, 23–24.

[10] John McPhee, *Basin and Range*, New York:Farrar, Straus and Giroux, 1981, pp.95-100.

[11] John McPhee, *Basin and Range*, pp.113-114.

[12] (in Russian) *Great Soviet Encyclopedia* (3rd ed.). Moscow: Sovetskaya Enciklopediya. 1974. vol. 16, p. 50.

[13] Geologic Time Scale (http://www.enchantedlearning.com/subjects/Geologictime.html)

[14] How the discovery of geologic time changed our view of the world (http://www.bris.ac.uk/news/2007/5609.html), Bristol University

[15] Felix M. Gradstein, James G. Ogg, Alan G. Smith (Editors); *A Geologic Time Scale 2004*, Cambridge University Press, 2005, (ISBN 0-521-78673-8)

[16] Cox & Richard, *A formal model for the geologic time scale and global stratotype section and point, compatible with geospatial information transfer standards*, Geosphere, volume 1, pp 119-137 (http://www.gsajournals.org/perlserv/?request=get-abstract&doi=10.1130/ GES00022.1), Geological Society of America, 2005

[17] International Commission on Stratigraphy. "International Stratigraphic Chart" (http://www.stratigraphy.org/upload/ISChart2009.pdf). . Retrieved 2009-09-25.

[18] Paleontologists often refer to faunal stages rather than geologic (geological) periods. The stage nomenclature is quite complex. For an excellent time-ordered list of faunal stages, see "The Paleobiology Database" (http://flatpebble.nceas.ucsb.edu/cgi-bin/bridge. pl?action=startScale). . Retrieved 2006-03-19.

[19] Dates are slightly uncertain with differences of a few percent between various sources being common. This is largely due to uncertainties in radiometric dating and the problem that deposits suitable for radiometric dating seldom occur exactly at the places in the geologic column where they would be most useful. The dates and errors quoted above are according to the International Commission on Stratigraphy 2004 time scale. Dates labeled with a * indicate boundaries where a Global Boundary Stratotype Section and Point has been internationally agreed upon: see List of Global Boundary Stratotype Sections and Points for a complete list.

[20] Historically, the Cenozoic has been divided up into the Quaternary and Tertiary sub-eras, as well as the Neogene and Paleogene periods. The 2009 version of the ICS time chart (http://www.stratigraphy.org/upload/ISChart2009.pdf) recognizes a slightly extended Quaternary as well as the Paleogene and a truncated Neogene, the Tertiary having been demoted to informal status.

[21] For more information on this, see the following articles: Earth's atmosphere, carbon dioxide, Carbon dioxide in the Earth's atmosphere, global warming, climate change, Image:Phanerozoic_Carbon_Dioxide.png, Image:65 Myr Climate Change.png, Image:Five Myr Climate Change.png, and Template:DF temperature

[22] The start time for the Holocene epoch is here given as 11,430 years ago ± 130 years (that is, between 9610 BC-9560 BC and 9350 BC-9300 BC). For further discussion of the dating of this epoch, see Holocene.

[23] In North America, the Carboniferous is subdivided into Mississippian and Pennsylvanian Periods.

[24] The Precambrian is also known as Cryptozoic.

[25] The Proterozoic, Archean and Hadean are often collectively referred to as the Precambrian Time or sometimes, also the Cryptozoic.

[26] Defined by absolute age (Global Standard Stratigraphic Age).

[27] The age of the oldest measurable craton, or continental crust, is dated to 3600–3800 Ma

[28] Though commonly used, the Hadean is not a formal eon and no lower bound for the Archean and Eoarchean have been agreed upon. The Hadean has also sometimes been called the Priscoan or the Azoic. Sometimes, the Hadean can be found to be subdivided according to the lunar geologic time scale. These eras include the Cryptic and Basin Groups (which are subdivisions of the Pre-Nectarian era), Nectarian, and Early Imbrian units.

[29] These unit names were taken from the Lunar geologic timescale and refer to geologic events that did not occur on Earth. Their use for Earth geology is unofficial.

[30] Bowring, Samuel A.; Williams, Ian S. (1999). "Priscoan (4.00–4.03 Ga) orthogneisses from northwestern Canada". *Contributions to Mineralogy and Petrology* **134** (1): 3. Bibcode 1999CoMP..134....3B. doi:10.1007/s004100050465. The oldest rock on Earth is the Acasta Gneiss, and it dates to 4.03 Ga, located in the Northwest Territories of Canada.

[31] Geology.wisc.edu (http://www.geology.wisc.edu/~valley/zircons/Wilde2001Nature.pdf)

External links

- NASA: Geologic Time (http://rst.gsfc.nasa.gov/Sect2/Sect2_1b.html)
- GSA: Geologic Time Scale (http://www.geosociety.org/science/timescale/timescl.htm)
- British Geological Survey: Geological Timechart (http://www.bgs.ac.uk/Education/britstrat/home.html)
- GeoWhen Database (http://www.stratigraphy.org/geowhen/)
- International Commission on Stratigraphy Time Scale (http://www.stratigraphy.org/gssp.htm)
- Chronos.org (http://www.chronos.org)
- National Museum of Natural History - Geologic Time (http://www.nmnh.si.edu/paleo/geotime/index.htm)
- SeeGrid: Geological Time Systems (https://www.seegrid.csiro.au/twiki/bin/view/CGIModel/ GeologicTime) Information model for the geologic time scale

- Exploring Time (http://exploringtime.org/?page=segments) from Planck Time to the lifespan of the universe
- Episodes.org (http://www.episodes.org/backissues/272-yasuo/Time Scale.pdf), Gradstein, Felix M. et al. (2004) *A new Geologic Time Scale, with special reference to Precambrian and Neogene*, Episodes, Vol. 27, no. 2 June 2004 (pdf)
- Lane, Alfred C,, and Marble, John Putman 1937. Report of the Committee on the measurement of geologic time (http://books.google.ca/books?id=ckIrAAAAYAAJ&printsec=toc&source=gbs_summary_r& cad=0#PPP1,M1)
- Lessons for Children on Geologic Time (http://www.newsciencelessons.com/geology_lesson_plans.html)
- Deep Time - A History of the Earth : Interactive Infographic (http://deeptime.info)

ltg:Geologiskuo laika skala

Kimmeridge Clay

Kimmeridge Clay Stratigraphic range: Upper Jurassic	
Type	Geological formation

The **Kimmeridge Clay Formation** is a sedimentary deposit of fossiliferous marine clay which is of Jurassic age. It occurs in Europe.

Kimmeridge Clay is arguably the most economically important unit of rocks in the whole of Europe, being the major source rock for oil fields in the North Sea hydrocarbon province. It has distinctive physical properties, log responses, and palynological signature.

It is named after the village of Kimmeridge on the Dorset coast of England, where it is well exposed anf forms part of the Jurassic Coast World Heritage Site. It exists across England, in a band stretching from Dorset in the south-west, north-east to East Anglia.

The Humber Bridge's foundations are in the Kimmeridge Clay deposits under the Humber estuary.

The fossil fauna of the Kimmeridge Clay includes a reptile fauna of turtles, crocodiles, sauropods, plesiosaurs, pliosaurs and ichthyosaurs, as well as a number of invertebrate species.

Vertebrate fauna

[1]

Ornithischians

Indeterminate nodosaurid remains have been found in Wiltshire, England.[1] Indeterminate stegosaurid remains have been found in Dorset and Wiltshire, England.[1]

Color key

Taxon	Reclassified taxon	Taxon falsely reported as present	Discredited taxon	Ichnotaxon	Ootaxon	Morphotaxon

Notes

Uncertain or tentative data are in small text; ~~crossed out~~ data are discredited.

Genus	Species	Location	Member	Abundance	Notes	Images

Genus	Species	Location	Stratigraphic position	Material	Notes	Images
Bugenasaura[1]	Indeterminate[1]	• Dorset[1]			Kimmeridge clay remains previously identified as belonging to *Bugenasaura* are now regarded as the remains of an indeterminate euornithopod.[1] However, the genus is also defunct now anyway, as scientists have determined it to be a junior synonym of *Thescelosaurus*.	*Dacentrurus*
Cumnoria[1]	*C. prestwichii*[1]	• Oxfordshire[1]		"Fragmentary skull and skeleton."[2]		
Dacentrurus[1]	*D. armatus*[1]	• Cambridgeshire[1] • Dorset[1] • Wiltshire[1]			Wiltshire remains include specimens previously referred to *Omosaurus armatus* and *O. hastiger*.[1]	
Omosaurus[1]	*O. armatus*[1]	• Wiltshire[1]			Reclassifed as *Dacentrurus armatus* because the generic name *Omosaurus* was preoccupied.[1]	
	O. hastiger[1]	• Wiltshire[1]				

Saurischians

Indeterminate ornithomimmid remains have been found in Dorset, England.[1] An undescribed theropod genus was found in Dorset.[1]

Color key

Taxon	Reclassified taxon	Taxon falsely reported as present	Discredited taxon	Ichnotaxon	Ootaxon	Morphotaxon

Notes

Uncertain or tentative data are in small text; crossed-out data are discredited.

Genus	Species	Location	Stratigraphic position	Material	Notes	Images

Bothriospondylus[1]	*B. suffosus*[1]	• Wiltshire[1]		"[Seven] dorsal and sacral centra."[3]	
Cetiosaurus[1]	*C. humerocristatus*[1]	• Dorset[1]			Now *Duriatitan*.[4]
	Indeterminate[1]	• Oxfordshire[1]			Remains previously referred to an indeterminate species of *Cetiosaurus* are now regarded as indeterminate sauropod material.[1]
Duriatitan	*D. humerocristatus*	• Dorset[1]		Humerus[5]	A titanosauriform[4]
Gigantosaurus[1]	*G. megalonyx*[1]	• Cambridgeshire[1]			Remains previously referred to *Gigantosaurus megalonyx* are now regarded as indeterminate sauropod material.[1]
Ischyrosaurus[1]	*I. manseli*[1]	• Oxfordshire[1]		"Humerus."[6]	Remains previously referred to *Ischyrosaurus manseli* are now regarded as indeterminate sauropod material.[1]
Megalosaurus[1]	*M. insignis*[1]	• Wiltshire[1]			Remains previously referred tentatively to *Megalosaurus insignis* are now regarded as indeterminate theropod material.[1]
	Indeterminate[1]	• Dorset[1]			Remains previously referred to an indeterminate species of *Megalosaurus* are now regarded as indeterminate theropod material.[1]
Ornithopsis[1]	*O. leedsi*[1]	• Oxfordshire[1]		"Caudal vertebrae, pelvis."[6]	Remains previously referred to *Ornithopsis leedsi* are now regarded as indeterminate sauropod material.[1]
	Indeterminate	• Cambridgeshire[1] • Norfolk[1]			Remains previously attributed to one or more indeterminate species of *Ornithopsis* are now regarded as possible indeterminate sauropod material.[1]

Invertebrates

The invertebrate fauna of the Kimmeridge Clay includes[7] [8] :

- Mollusca:
    - *Cardium striatulum*
    - *Ostrea deltoidea*
    - *Gryphaea (Exogyra) virgula*
    - The ammonite aptychus known as *"Trigonellites latus"*
    - *Belemnotheutis*

See also

- List of dinosaur-bearing rock formations

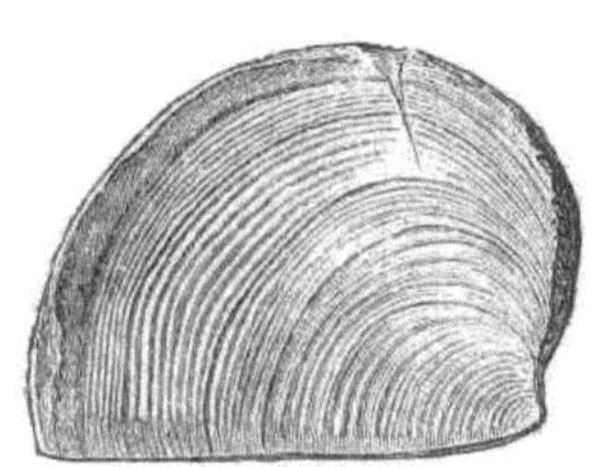

An aptychus with the name "Trigonellites latus",
from the Kimmeridge Clay Formation

References

[1] Weishampel, David B; et al. (2004). "Dinosaur distribution (Late Jurassic, Europe)." In: Weishampel, David B.; Dodson, Peter; and
     Osmólska, Halszka (eds.): The Dinosauria, 2nd, Berkeley: University of California Press. Pp. 545–549. ISBN 0-520-24209-2.

[2] "Table 19.1," in Weishampel, et al. (2004). Page 415.

[3] "Table 13.1," in Weishampel, et al. (2004). Page 270.

[4] Paul M. Barrett, Roger B.J. Benson and Paul Upchurch (2010). "Dinosaurs of Dorset: Part II, the sauropod dinosaurs (Saurischia, Sauropoda)
     with additional comments on the theropods". *Proceedings of the Dorset Natural History and Archaeological Society* **131**: 113–126.

[5] "Table 13.1," in Weishampel, et al. (2004). Page 267.

[6] "Table 13.1," in Weishampel, et al. (2004). Page 271.

[7] http://www.fullbooks.com/The-Student-s-Elements-of-Geology7.html The Student's Elements of Geology by Sir Charles Lyell Part 7 out
     of 14 accessed 13 February 2009.

[8] Wignall, Paul B. (1990). "Benthic palaeoecology of the late Jurassic Kimmeridge Clay of England" (http://rogov.zwz.ru/Wignall,
     1990_Bentic palaeoecology_Kimmeridge clay.pdf). *Special Papers in Palaeontology* (The Palaeontological Association, London) **43**.
     ISBN 9780901702425. . Retrieved February 8, 2011.

George Rolleston

George Rolleston MA MD FRCP FRS (30 July 1829 – 16 June 1881) was an English physician and zoologist. He was the first Linacre Professor of Anatomy and Physiology to be appointed at the University of Oxford, a post he held from 1860 until his death in 1881. Rolleston, a friend and protégé of Thomas Henry Huxley, was an evolutionary biologist.[1]

Life

Rolleston was born at Maltby Hall, near Rotherham, Yorkshire, England.[1] His parents were probably Rev. George Rolleston (rector and squire of Maltby) and Anne Nettleship; and his brother probably William Rolleston, later of New Zealand.[2]

Rolleston was educated at Queen Elizabeth's Grammar School, Gainsborough;[3] Sheffield Collegiate School; Pembroke College, Oxford and St Bartholomew's Hospital, London. He qualified with the degrees of BA (1850, 1st Class), MA and MD. The same year he

George Rolleston

entered Pembroke College, Oxford, and took a First Class in Classics. After qualifying as a physician, Rolleston became a Fellow of Pembroke College in 1851, holding posts at the British Civil Hospital, Smyrna (during Crimean War), and Assistant Physician, Children's Hospital, London (1857). Gradually he became more interested in zoology, and spent the rest of his career as a zoologist, and on the human sciences. His research included comparative anatomy, physiology, zoology, archaeology and anthropology. In 1860, he was elected to the newly founded Linacre Professorship of Anatomy and Physiology, which he held to the time of his death. He became FRCP in 1859, was elected Fellow of the Royal Society on 5 June 1862, and Fellow of Merton College, Oxford, in 1872. He was a member of the Council of the Oxford University, its representative in the General Medical Council, and also an active member of the Oxford Local Board.[1]

In 1861, Rolleston married Grace, daughter of John Davy FRS and niece of Sir Humphry Davy; they had seven children.[1] He died of uraemic convulsions in Oxford in 1881 and is buried in Holywell Cemetery, Oxford. His anthropological archive came to the Ashmolean Museum, along with the archaeological material resulting from his excavations.

One of his sons was Sir Humphry Rolleston, an eminent physician himself.

Career

As a zoologist, Rolleston was a protégé of Thomas Henry Huxley, and took part in both of the critical sessions at the 1860 British Association meeting in Oxford.[4] Rolleston was one of the organisers for the meeting: he arranged for Huxley to stay at Christ Church during the meeting, and to have a crocodile skull in Huxley's room for study. Huxley was instrumental in Rolleston's appointment to the Linacre chair that very year, backing him against Owen's candidate. Rolleston wrote him a 'you'll never regret this' letter.[5]

Bust of George Rolleston in the Oxford University Museum

As an expert on the brain, Rolleston was present on the Thursday, when Huxley denied Owen's claim that human brain had parts that apes did not, and again on the Saturday for the debate on Darwin, where his opponent was Bishop Samuel Wilberforce. Rolleston was an Anglican, but a liberal in his religious beliefs, as was Huxley's other supporter in the brain debate, William Henry Flower. Huxley organised his FRS, as he did for Flower; and the two men acted as liaison between the X-Club and the Royal Society.[6] Rolleston remarked later that whenever he lectured on evolution, he was asked 'Was I an atheist or a Unitarian?' and some of Huxley's attacks on the Old Testament did cause him anguish.[7] [8]

Rolleston was so identified with Huxley at this time that he appeared as one of 'Tom Huxley's low set' in the ironical skit *Report of a sad case recently tried before the Lord Mayor, Owen versus Huxley* (publ. George Pycraft 1863) as 'Charlie Darwin the pigeon-fancier and Rollstone' cheer on their barrow-boy. This vivid broadsheet was certainly well informed: it mentions Owen's disgraceful maltreatment of Gideon Mantell.

See also

- William Henry Flower#London: transfer to zoology
- Alfred Newton#Reception of the Origin of Species

Publications

- 1870. *Forms of animal life: a manual of comparative anatomy*. Oxford.
- 1888. *Scientific papers and addresses* [9]. 2 vols, Oxford.

Footnotes

[1] "Obituary Notices of Fellows Deceased" (http://www.archive.org/details/philtrans00812237). *Proceedings of the Royal Society of London* **33**: i–xxvii. 1881. doi:10.1098/rspl.1881.0061. .
[2] same birth place; uncommon name; limited number of educated families in a small location
[3] now Queen Elizabeth's High School, Gainsborough
[4] Desmond, pp. 274–6, 280–1
[5] Rolleston to Huxley, Huxley Papers (Imperial College), HP 25.142, 148, 150.
[6] Desmond, pp. 306, 329
[7] Rolleston to Huxley 1 Jan 1865 HP 25.167
[8] Desmond, pp. 331–2
[9] http://books.google.com/books?id=NykaAAAAYAAJ&pg=PR10&dq=George+Rolleston&as_brr=1

References

- Desmond, Adrian 1994. *Huxley*: vol 1 *The Devil's Disciple*, London: Michael Joseph ISBN 0718136411
- Bulloch's Roll; DNB; DSB

Article Sources and Contributors

Iguanodont *Source*: http://en.wikipedia.org/w/index.php?title=Iguanodont *Contributors*: ALE!, Anaxial, Badseed, Ballista, CALR, CFLeon, Cephal-odd, Darkwind, Dinoguy2, Elmo12456, Enlil Ninlil, Evanphoto, Firsfron, FunkMonk, Gabriel Solo, Grand Moff Brian, Hooperbloob, Hoseumou, Hurricane111, J. Spencer, Jan.Kamenicek, Jerkov, Johan Moberg, Jrockley, Kumioko, MWAK, Mgiganteus1, Mitch Sutton, Rjwilmsi, Rnnsh, Stho002, Tim1357, Velociraptor888, 19 anonymous edits

Cumnoria *Source*: http://en.wikipedia.org/w/index.php?title=Cumnoria *Contributors*: FeatherPluma, Firsfron, FunkMonk, Hamamelis, J. Spencer, MWAK, Peter coxhead, Rnnsh, 1 anonymous edits

Dinosaur *Source*: http://en.wikipedia.org/w/index.php?title=Dinosaur *Contributors*: *Kat*, *drew, 1-is-blue, 10metreh, 10outof10die, 20075896cu, 2038432048234, 216.175.251.xxx, 2468whodowe, 28421u2232nfenfcenc, 3dnatureguy, 4444hhhh, 5dollabill, 83d40m, 96.149, A Macedonian, AMC0712, Aadal, Aaron Schulz, Aaron Walden, Abigail-II, Abyssal, Academic Challenger, Aceofelves, Adam78, Additional110, Adrian J. Hunter, Againme, Ahoerstemeier, Ajp2000, Albertonykus, Ale jrb, Aleron235, Alex Bakharev, AlexKepler, Aliter, Alpha3103, AmberUTS, Amberrock, Amicon, Amillar, AmyRowe, AnakngAraw, Ancheta Wis, Andre Engels, AndreNatas, Andrew Levine, AndrewHarvey4, Andrewlp1991, Andrewpmk, Andries, Andy, Andycjp, Angado keitho, Anger22, Anonymous editor, Antandrus, Anthony Appleyard, Aquishix, Aragupta, Archangel x105, Archfiendweazal, ArdcoreSoxx, ArielGold, Arpingstone, Art LaPella, ArthurWeasley, Artisticmerit, Aryeztur, Ashcraft, AssistantX, Astro$01, AstroHurricane001, Astropithicus, Athf1234, Attilios, Atulsnischal, Atun, Aunt Entropy, Autodidactyl, Ayrton Prost, Azhyd, B&W Anime Fan, Blatv, BAFAphenom7, BD2412, BWCNY, Babylittletree, Ballista, Bambadil, Banuthev, BartekChom, Basilicofresco, Bassem18, Bbatsell, Bcollecutt, Begoon, Ben Hateva, Ben Jos, Ben m, Benne, Bensaccount, Bento00, Bergsten, Best Known, Bevo, BhaiSaab, Bhumiya, Bidabadi, Bignole, Bill the Duck, Black Kite, Blackicehorizon, Bloodredrover, Blue520, Bob the Wikipedian, Bobber1, Bobblewik, Bobo192, Bodnotbod, Bomac, Bonesiii, BorgHunter, Brad101, BrendanRyan, Brendanconway, Brhoops1, Brian0918, Brighterorange, Brnstormersfan, Bryan Derksen, Bubalana, Buchanan-Hermit, Buckshot128, Buickid, Bulltastrophy, Butros, Buwayahman, Bwilliams, Bwmodular, C4h5n3o, CBDunkerson, CJLL Wright, Caltas, Camerong, Can't sleep, clown will eat me, Canadian-Bacon, CanadianCaesar, Canderson7, Captain Mar vell, Captain02, Captainbeefart, Casliber, Cateybug, Cavrdg, Cbdorsett, Cbogart2, Cdc, Cedrus-Libani, Celestianpower, Cenarium, Centrx, Cephal-odd, Chanting Fox, Chaos, Chowbok, Chris 73, Chris Capoccia, ChrisGriswold, Christian List, Christopherlin, Chuck Marean, Cimp3, Circeus, Cleverclogs06, ClockworkSoul, ClockworkTroll, Cmanoli, Colin stuart, Colonies Chris, CommonsDelinker, Comp25, ConfuciusOrnis, Conkyworm, Connormah, Conti, Conversion script, Cosmo1, Cpandtle, Crag, Craigsjones, Crot buer, Crotalus horridus, Csaw, Curps, Cutter, Cvene64, Cwruidth, Cyberevil, CyclePat, Cyclopia, D, D-Notice, DARTH SIDIOUS 2, DJ Clayworth, DLJessup, DMacks, DSedrez, DVD R W, DabMachine, Dan D. Ric, Danarmak, Daniel J. Leivick, Daniel.Cardenas, DanielCD, Danny, DarkMasterBob, Darwin's Waiting Line, Dave souza, David Fuchs, David Latapie, David Little, Dbachmann, Debivort, Deckiller, Decltype, Degina2008, DeinonychusDinosaur999, Dendodge, Denovoid, Derek Yoda's friend, Deror avi, Desmay, Devotchka, Diberri, DietCaffeineFreeCherryCoke, Digitalme, Dinoguy2, Dinohunter, Dinosauria-Freak, Dinosaurs12345, Discordance, Discospinster, Diucón, Diyako, Dj Capricorn, Dlohcierekim's sock, Doc Tropics, DoctorLarry, Dolovis, Dom Kaos, Domminico, Dr.Bastedo, Dracontes, Dragon Helm, DragonflySixtyseven, Draicone, Drakesiphon, DreamOfMirrors, Dromaeosaur Dude, DropDeadGorgias, Dropzink, Drunkenmonkey, Dudo2, Dysepsion, Dysprosia, ESkog, EWS23, Ed Cormany, Eeeasterbunny, ElTyrant, ElZorroLivesForever, Eleassar, Elembis, Elkester, Elliskev, Elmosdapimp, Eloquence, Eluchil404, Elvarg, Encephalon, Ensrifraff, Epbr123, Eras-mus, Espi, Estel, Evercat, Everyking, Everything212, Evice, EvilZak, Ezra Wax, FF2010, Fama Clamosa, Fan-1967, Farsight001, Fastfission, Fastzander, Fat kidddd, Favyan, Fconaway, Felix Dance, Fenevad, FiGGy-06, Finalius, Firsfron, Flamingspinach, Florentino floro, Fotoguzzi, Freakofnurture, Frecklefoot, Fredrik, Frencheigh, Frosty0814snowman, Fruge, FunkMonk, Funnybunny, Futuremyst, Fuzheado, Fyunck(click), GDallimore, GGordonWorleyIII, GHe, GSGSGSG, GSlicer, Gabbe, Gaga654, Gaia5074Q, Gaius Cornelius, Galassi, Garion96, GarrettD, Garry Denke, Gavin the Chosen, Gazza69, Gazzster, Gene Nygaard, Geologyguy, Gfoley4, Gfuzz, Giftlite, Gilgamesh, Gilliam, Gimlifangirl, Gimmetrow, Givenarmy, Glenn, Globalmuseum, Gloomy Storm, Gnomehead, Gnostrat, Goggle archives, Gogo Dodo, GrAveTzT, GraemeL, Grafen, Graham87, GrahamN, Graminophile, Greanak, GreatWhiteNortherner, Greebo cat, GregRustFan, Gregalodon, Grundle2600, Grutter, Gsklee, Gtrmp, Guambomb, Guybrush, Gwernol, H0merg0mez, H2g2bob, HKMP5A2, HMallison, Hadal, Haeleth, Hall Monitor, Hamiltondaniel, Hans Dunkelberg, Happyguy123, Happyguy32, Happysailor, Harley peters, Harold137, Harrisonjason, Hastyo, Heial123456, Henryisamazingyeh, Henzel, Hermione1980, Herp Derp, Heteren, HiDrNick, Hibernian, Hiddensonyvaio, Hires an editor, HistoryBA, Hmacmil2, Hmains, Hobartimus, HonoraryMix, HoodedMan, Hordaland, Hoseumou, Hottentot, HunterX, Hut 8.5, I 0wn in rs, I teh yuh, IVAN3MAN, Ialsoagree, Icalanise, Ikh, Improv, Indosauros, Infraredeclipse, Innotata, Inventedgravity, InverseHypercube, Invincible Ninja, Iota, Irbnspgn, Irixman, Irvin calicut, Islandbaygardener, Isomorphic, Its me, all feakin me, Itsmine, J. Spencer, JCarriker, JForget, JK43, JOSHKIN, JRLivesey, JW1805, Jacek Kendysz, Jacob.jose, Jafeluv, Jameswilson, JamieSwan1993, Jamotive, Jannex, Jaranda, JarlaxleArtemis, Jason Potter, Jaxl, JayC, Jayen466, Jc-S0CO, Jchap, Jckrull, Jedi of redwall, JeffStickney, JeffTL, Jefffire, Jens Lallensack, JeremyA, Jerkov, JerryOrr, Jfg284, Jiang, Jidanni, Jiddisch, Jim62sch, Jimp, JinJian, Jipcy, Jiy, Jkbjlfk, Jklamm, Jklin, Jmv2009, Jmw0000, Jni, JoeSmack, Joeyrockyhorror, Johann Wolfgang, John Doe or Jane Doe, John Reid, John of Reading, John.Conway, JohnWayne, Jokaer, Jokkel, Jonas Poole, Jorjepot, Josh Grosse, JoshuaZ, Jossi, Joyous!, Jpcoles, Jpo, Jrsnbarn, Jtkiefer, JudahH, Julia W, Jumacdon, Jumbo Snails, Jusdafax, JustinWick, Jwissick, Jyril, Jóna Þórunn, K, Kaijucole, Kandar, Kaoruchan, Karam.Anthony.K, Karimarie, Kazooieman, Kazvorpal, Kbdank71, Kchishol1970, Kdakin, Kent Wang, Kevin B12, Kevindavid, Kglavin, Khoikhoi, Kickaha, Kilgore Sprout, Killdevil, KillerChihuahua, Kilo-Lima, Kintaro, Kipala, Kirill Lokshin, Kirkseeray, Kit Foxtrot, Kleopatra, Klonimus, Kmccoy, KnightRider, Knowledge Examiner, Knowledge Seeker, KnowledgeOfSelf, Knutux, Koliri, Kotengu, Kozuch, KrAtul, KrakatoaKatie, KritonK, Kronermark, Kubigula, Kuru, Kurykh, Kwamikagami, Kwiki, Kyle Barbour, Kyle sb, L Kensington, La goutte de pluie, Lacrimosus, Lanasa, LandonChastelhassy, Landroo, Latka, Lavateraguy, Lbaum, LeeWilson, Leeteditmastorz, LeonWhite, LeonardoRob0t, Lepticitidium, Lepton6, Leroyinc, Lexi Marie, Lightmouse, Ligulem, Like omg its karma, LilHelpa, Liridon, Lizhia, Llull, Lord Seth, LordAmeth, Lot49a, Lowellian, Ludwigs2, Lugnuts, Luna Santin, Lupin, Luuva, M C Y 1008, MCBOB, MER-C, MPF, MSGJ, MWAK, MZMcBride, Maastrictian, Mac Dreamstate, MacTire02, Mackinaw, Madchester, Madtast, Magister Mathematicae, Mahanga, Majungatholus, MakeChooChooGoNow, MaIK`k, Malo, Mandyjb, Manufracture, MarkBuckles, MarkGallagher, MarkSweep, Markeer, Markjreed, MarvintheParanoidAndroid, Massimo Macconi, Master of Puppets, MasterOfHisOwnDomain, Masterv88, Matt Crypto, Matt McIrvin, Matt me, Matt1314, MattDP, MattKingston, Mattbrown04, MatthewAJYD, Mausy5043, Mav, Max Schwarz, Maximus Rex, Maxthefax123, Mazza024, Mckay history, Mdd4696, Mebden, MechBrowman, Medeis, Mediatenz, Meggar, Melaen, Melesse, Menchi, Mephistophelian, Meridas, Mermaid from the Baltic Sea, Mgglaser, Mgiganteus1, Mgumn, Michael Devore, Michael Zimmermann, Michael-david Reisman, Michaelmas1957, Michaelritchie200, Michaelrogerson, Mike Dillon, Mike Rosoft, MikeLynch, MilkshakeMan246, Millahna, Minna Sora no Shita, Misza13, Mjbaldwin, Mkdw, Mn extra, Mo0, Mokele, Mokwella, Mollwollfumble, Monicasdude, Moon&Nature, Moreschi, Mormegil, Morrad, Moshe Constantine Hassan Al-Silverburg, Movementarian, Movieresearch, Mr w, Mrdallaway, Mrmanager07, Mschel, Mschlindwein, MuDavid, Murtasa, My76Strat, Myfretsareonfire, Mysid, Mzmadmike, Nakon, Name Theft Victim, Narayanese, Natgoo, NatureA16, NawlinWiki, Nbarth, Neale Monks, Nemokid1220, Netsnipe, Neutrality, Nick Number, Nick UA, Nick125, Niczar, Nikodemos, NinjaCharlie, Ninjatacoshell, No substitute for you, Noah Lawrence, Nodosaurus, Noirish, Not an anon anymore, Nrg800, NuclearWarfare, Nuno Tavares, Nyy2135525, Oblivious, Obradovic Goran, Ocaasi, Ohnoitsjamie, Okapi, Old Father Time, OldakQuill, Oldshonen, Ollivier, Omicronpersei8, Orangemarlin, OrgasGirl, Orie1061, Otolemur crassicaudatus, Ours18, Ovons, Owen, PDH, PJS102, Palica, Pamri, Panzuriel, Paranoid, Parsa, Patrickncollin1, Patricksmullen, Paul Erik, Paul Magnussen, Paul Stansifer, Pcb21, Peachsncream, Pegship, PenguinJockey, Penguinpablo, Perstar, Peruvianllama, Peter Isotalo, Petri Krohn, Petter Bøckman, Pfhreak, Pgk, Phaedriel, Phatom87, Phifer784, Phil Boswell, Phil1988, Philcha, Philg88, Phlebas, Phoenix79, PiCo, PiMaster3, PiRSquared17, Pickelbarrel, Picone, Pie4all88, Piloscia, Pilotguy, Pinethicket, Pjrich, Pleebloo, Poison the Well, Pomeapplepome, Porqin, Portillo, Predator, PrestonH, Pretty Green, Primate, Pufnstuf, Quadell, Quantum bird, Quercusrobur, R. Baley, R.D.H. (Ghost In The Machine), RG2, RPlunk2853, RWFanMS, Rahimbakhsh, Ramdrake, Rampart, Randomblue, Raptorfact, Raul654, Raven in Orbit, Rdsmith4, Reach Out to the Truth, Rebecca Bangay, RedSpruce, RedWolf, Reddish, Redthoreau, Reedy, RetiredUser2, Retran, Revansrangers, Revolución, RexNL, Riana, Rich Farmbrough, Richard Keatinge, Ringsjöodjuren, Rintrah, Ritchy, Rje, Rjwilmsi, Rm w a vu, Roaribeadino, Rob117, RobertG, RoboAction, Rocastelo, Rogor, Roke, Roleplayer, Rossnixon, Rothorpe, Roy da Vinci, RoyBoy, Royalguard11, Rsm99833, Rufous-crowned Sparrow, Russell C. Sibley, RyanGerbil10, Ryanaxp, Ryanfrm, Ryanmcg2006, Ryoung122, Ryulong, Ryuuseipro, SFH, Sai2020, SallyForth123, Sam Hocevar, Sampi, Samsara, Sango123, Sardanaphalus, Saucerottia, Scary dragon atop the hill, Sceptre, SchuminWeb, Scm83x, Screwwithu99, Scroner, Scruffian, Seaphoto, Seb az86556, Sengkang, Serendipodous, ShadowMan1od, Shadowjams, Shagie, Shaki997, Shanel, Shanes, Shantavira, Sheeana, Sheep81, Sheridan, Shizhao, Shoaler, Siblin, Silverwood, SimonP, SineWave, Siriusblack902, Sjakkalle, Sjorford, Skizzik, SlimVirgin, Smith609, Sneaky Oviraptor18, Snowolf, SoCalSuperEagle, Soap, Someone56, Spacemanjohn, Spangineer, Sparky the Seventh Chaos, SparrowsWing, Spawn Man, Spinodontosaurus, Spinosaur, Spinxy, SpookyMulder, Sports are my world, Spotty11222, Springnuts, Sputnikcccp, Srleffler, Srnec, Sstty4, Starexplorer, Stbalbach, Steinbach, Steinsky, Stemonitis, Stephan Schulz, Stephen G. Brown, Stephen Gilbert, Stephen e nelson, Stephenb, Steven Walling, Steveoc 86, Stickman on a stick, StoptheDatabaseState, Streona, Stubblyhead, Studio34, Stuff8, Subidei, Sucatraps, SunnyDisp, SuperHamster, Superraptor, Supreme Deliciousness, SusanLarson, Swatjester, Systolic, T.lancer, Tab439, Tannin, Tapir Terrific, TashTish, Tbhotch, Tdlugolecki, TeaDrinker, Tekana, Temtem, TestPilot, Thanatosimii, The Librarian At Terminus, The Man in Question, The Rambling Man, The Ronin, The Singing Badger, The Thing That Should Not Be, The Utahraptor, TheBigJabroni, TheCryingofLot49, TheDevilsTrill, TheFixer000, TheJC, TheKMan, TheProject, TheTruthiness, Theda, ThinkBlue, This user has left wikipedia, Thumperward, Thunderhead, Tide rolls, TigerShark, TimBentley, TimVickers, Timotheus Canens, Tiptoety, Titanium Dragon, Titoxd, Tjs003, Tjunier, Tktktk, Tmangray, Tmoney792, Tmopkisn, Tobby72, Toffelginkgo, Tom-, TomStar81, Tomalak geretkal, Tommulliner, Tommy2010, Tomsega, Tone, Tony Fox, TonyClarke, Tosoft, Tothebarricades.tk, Tpbradbury, Train2104, Traveler100, Travmitz, TreasuryTag, Trekphiler, Trevor Andersen, Trevor MacInnis, Triceratops, Trojanhero, Troodon58, Tropylium, Trovatore, Truthteller, Trödel, Tsunamicarlos, Tsunomaru, Tuckwai, Tuohirulla, Tyrannosaurus roll, Tyrol5, Unclerico89, UndeadMonkey, Undress 006, UtherSRG, VAcharon, Vague Rant, Vanished User 0001, Vanished user, Vanished user 03, VanishedUser314159, Vargklo, VegaDark, Veledan, Vendetta21, Vera.tetrix, Vessss, Vikreykja, Vina, Violetriga, Vsmith, W33nie, WAS 4.250, WaffleMafia, Wafulz, Waggers, Waker913, Warfwar3, Wars, WarthogDemon, Wavelength, Wayward, Wbroun, WereSpielChequers, Weregerbil, Wesley J M, Westvoja, Wetman, Whatiguana, WhatThree16, Whimemsz, Whooper, Wiggles649, Wik, Wiki alf, Wikipages, Wikipediarules2221, Wikivandalizer, Wiknox, WillNess, William Avery, Wilson44691, Wimt, Winston365, Wkey, WolfmanSF, Woohookitty, Worldtraveller, Wowaconia, Xaosflux, Xerxes314, Yamamoto Ichiro, Yasharelliiin, Yes234, Yggdrasilsroot, Yooden, Yossarian, Yummifruitbat, Zaffo, Zaharous, Zanimum, ZeWrestler, Zfervent, Zharta, ZimZalaBim, Zoologyteacher, Zoso95, Zsinj, Zywxn, Ævar Arnfjörð Bjarmason, 1067 anonymous edits

Genus *Source*: http://en.wikipedia.org/w/index.php?title=Genus *Contributors*: 16@r, 5 albert square, A Macedonian, Abanima, Addionne, Aitias, Ajraddatz, Alansohn, Alex.muller, Alexei Kouprianov, AlexiusHoratius, Andre Engels, AndrewWTaylor, AnonMoos, Anonymous Dissident, Anthony Appleyard, Aranea Mortem, Arichnad, Art LaPella, AstroNomer, Athaler, AugPi, Austroungarika, Autonova, AxelBoldt, Baranxtu, Belfry, Benjamnjoel2, Berria, Berserkerz Crit, Berton, Betterusername, Bjankuloski06en, Bobblewik, Bobo192, Bongwarrior, Brian0918, Brianga, Brya, Bryan Derksen, Byron755, CWY2190, Cburnett, Christiangrann, Circeus, CrazyUkee, Cxz111, DMacks, Damnreds, Darth Panda, Daviddariusbijan, Dendrid, DerHexer, Don2k11, Dr CyCoe, Drew roy, Drmies, Dysmorodrepanis, ENeville, ERcheck, ESkog, El C, Eloquence, Emma1700, Enigmaman, Eog1916, Epbr123, Erebus Morgaine, Escape Orbit, Ethien, Firsfron,

Flewis, Flyset, Frankenpuppy, Fred Bauder, FreplySpang, FrozenPurpleCube, GB fan, GKoUtSo201, Galoubet, Gdr, GerardM, Giftlite, Gilliam, Glenn, GoEThe, GrahamBould, Graminophile, Grendelkhan, Grunny, Gwern, Hallows AG, Hamamelis, Harloshaply, Hayabusa future, Hbent, Heracles31, Hozro, Hyacinth, I got fived!, Ian Dalziel, Icairns, Ignitus, Iliketochangearticles, Innotata, Invertzoo, Iridescent, J.delanoy, JHolman, Jackfork, Jackjackjr, Japanese Searobin, JemGage, Jerzy, Joblax, John KB, Joseph Solis in Australia, Jossi, Jrockley, Julesd, Julian Mendez, JustAGal, Katimawan2005, Kazenteshi, Kchishol1970, Keilana, KeithTyler, Kesuari, Kimse, Kintaro, Kironide, Kjkolb, Kmsiever, Knutux, Komododragonfan16, Kubigula, Kukini, Kungfuadam, LOL, Lankiveil, Lantianer, Lerdsuwa, Lightmouse, Luckas Blade, MER-C, ML5, MPF, MWAK, Madhero88, Maelin, Maildej, Mani1, Marek69, Maristoddard, Marshman, Mauler90, Mav, Merovingian, Meteor2017, Mfwitten, Mgiganteus1, MichaelSeemann, Mikael Häggström, Mike Dillon, Mike Searson, Mikepenny01, MisfitToys, Mjster, Mmcknight4, Ms2ger, Muhaha, Mukkakukaku, Nakon, NewEnglandYankee, Nihiltres, Nivlak, Nnof002, Nsaa, Nutzmonkey, Oneiros, Orphan Wiki, Pageleft, Paul August, Paul-L, Pengo, Philip Trueman, PierreAbbat, Pinethicket, Pippu d'Angelo, Prashant.saxena, R.e.b., Ram-Man, RazorICE, Redaktor, RexNL, Rich Farmbrough, Richard New Forest, Richard0715, Rjwilmsi, Robert Foley, RobertG, Romanskolduns, Ronhjones, Rossco89, RoundAbout949, Rvollmert, S3000, SP-KP, Salamurai, Sarahcwilkins, Satellite9876, Scientizzle, Sciurinæ, Seahorseruler, Shirt58, Shulei-shulei, SimonP, Sluzzelin, Smith609, Snek01, Snowmanradio, Spiff, Srios653, Standproud1, Starks, Stemonitis, Super Mario, TUF-KAT, Taztouzi1, The Earwig, The Final Chronicler, Tide rolls, Tobias Bergemann, Tommy2010, TotoBaggins, Unara, Uncle G, Unyoyega, Uscurnow17, UtherSRG, Velho, Verbum Veritas, Victor falk, Vietbio, Vina, Vishnava, Vsmith, We.eter, Werts335, Whosyourjudas, Wilbozz, Williamb, Wisco, WolfmanSF, YellowMonkey, Yerpo, Yonwe, Yosri, YuKornilev, Zumbo, Саша Стефановић, 349 anonymous edits

Geologic time scale *Source*: http://en.wikipedia.org/w/index.php?title=Geologic_time_scale *Contributors*: 10metreh, 119, 1ForTheMoney, 2002champs, 216.192.77.xxx, 5 albert square, AbJ32, Abarry, Abce2, Acroterion, AdjustShift, Agathman, Alan clayton, Alansohn, Alaudo, Ale jrb, AlefZet, Alireza1984e, Altenmann, Amcguire, Amish 01, Andre Engels, Andres, Anktuman, Antandrus, Anthony Appleyard, Antman, Anturiaethwr, Arakunem, Athaenara, Attilios, Aunt Entropy, Autodidactyl, Awickert, AxelBoldt, Azcolvin429, Babakathy, Bacchus87, Bcasterline, Bdusel, Bejnar, Beland, Ben.c.roberts, Benny 919, Bergsten, Bignose, Bluezy, Bobo192, Bongwarrior, Brim, Bryan Derksen, Burwellian, CP2002, Can't sleep, clown will eat me, Canaima, CarolGray, Casito, Cedders, Cenarium, Cgingold, Chaos, Chasingsol, Chendy, Chermundy, Chris83, Chrisjohnston24, Cmapm, Codetiger, Colonies Chris, Cometstyles, Cst17, Daniel Olsen, Daniel.Cardenas, DanielCD, Daven200520, Deadly lava2006, Delldot, Dendodge, Devonta1213, Dia^, Dialectric, Dinoguy2, Dirrival, Dittaeva, Dixi, Dlohcierekim's sock, DocWatson42, Doctorwho123456, Don Kenney, Donarreiskoffer, Dozla, Dr Shorthair, Dr. Blofeld, Dr.Bastedo, Dragons flight, Dude1818, DuncanHill, Dwheeler, Dylan Lake, EagleFalconn, Ec5618, Edderso, EikwaR, Eleassar777, Enlil Ninlil, Epbr123, Erik Zachte, Eubulides, Eveilaje, EvilPizza, Evolauxia, FF2010, Fabartus, Fang Aili, Fieldday-sunday, Fradeve11, FreplySpang, GaborLajos, Garth 187, Gauss, GeoGreg, GeoWriter, Geoff.powers, Geologyguy, Ghelae, Giftlite, Gilgamesh, Gilliam, Glane23, Glenn, GraemeL, Graham87, Greeneto, Grutness, Gscshoyru, Gurch, Hagedis, Hairy Dude, Hans Dunkelberg, Hardwigg, Hardyplants, Hasek is the best, Hrafn, Hu12, Husond, IAMSka, II MusLiM HyBRiD II, Ihateduncan, Instinct, Iridescent, IvanLanin, Ixfd64, J04n, JAn Dudík, JCHYSK, JDspeeder1, Jackfork, Jagged 85, James.mcd.nz, Jamesmusik, JeffW, Jmundo, John Abbe, Jorfer, JoshHolloway, JoshuaZ, Jpeeling, Jrc494, Juliancolton, Just Another Dan, Jyril, Kalogeropoulos, Karenjc, Karl-Henner, Karmosin, Katanada, Kbdank71, Keephappyheyheyhey, Keilana, Keith Edkins, Kenny sh, Killian441, Kinaro, KingTT, Kjv31, Kleopatra, Knowledge Seeker, KnowledgeOfSelf, Koavf, Kontos, LaMenta3, Lejean2000, Leuqarte, Lightdarkness, Lightmouse, Likearock, LilHelpa, Longhornsg, Look2See1, Looteras, Looxix, Lotje, Lowellian, Lucid, Luna Santin, Lunokhod, Lusanaherandraton, M Alan Kazlev, MER-C, MKoltnow, MONGO, Mac, Marcok, Marioguy124, MartinDK, Martpol, Materialscientist, Matijap, MattieTK, Mav, Maziotis, McSly, Mccready, Mejor Los Indios, Mekong Bluesman, Mel 23, Menchi, Menswear, Mentifisto, Michael Hardy, MikaelFen, Mikenorton, Minna Sora no Shita, Mitchowen, Mmcannis, Modemac, Monado, Mordicai, Mountainguyspeaks, Moverton, Mwtoews, NJGW, Nachosan, Najro, Nakon, Nameless pl, NatureA16, Nburden, Nepenthes, Nessynessy1, Newsmare, NickW, Nicklott, Nicolae Coman, Ntsimp, Ntsloan, Nubula, NuclearWarfare, Nurg, Nwbeeson, Okay12, Ollj, Optimist on the run, Orange Suede Sofa, Orangemarlin, Orcoteuthis, Pannydragon, Parsa, Patrick, Paul August, Pengo, Persian Poet Gal, Peter b, Peter ja shaw, Pgan002, Pharaoh of the Wizards, Philip Trueman, Phoenix79, Pinethicket, Pinnerup, Piplz, Plasticty, Poetaris, Ptr123, Quantpole, QueenCake, Quercusrobur, Qwfp, Qxz, RB972, RJHall, Railsmart, Rcingham, Reformist, Rettetast, Rich Farmbrough, RickK, RingtailedFox, Rob Hooft, Rrostrom, RupertMillard, Rusty Cashman, SEWilco, Sairjohn, Sam Hocevar, Sam Korn, Samohyl Jan, Sango123, Scetoaux, Scientizzle, Scilit, Scipius, Seaphoto, Selyabd, Shantavira, Shawn in Montreal, Shythylacine, Signalhead, Sir48, Sjc, Slawojarek, SmartGuy, Smith609, Snalwibma, Somarinoa, Sophia geo, Spiffy sperry, Splash, Spotty11222, Springnuts, Stemonitis, Steven J. Anderson, Stevenmitchell, Stevey7788, Storm Rider, Stubblyhead, Suffusion of Yellow, Sushant gupta, Syncategoremata, THEN WHO WAS PHONE?, Tarquin, Tempodivalse, Thankyousir8, The Anome, The Obento Musubi, The Thing That Should Not Be, ThePointblank, Thisisverywierd3465, Thumperward, Tide rolls, Tim Shuba, TimVickers, Timothy Clemans, Tobby72, Toddst1, Triona, Triwbe, Tsogo3, Twooars, Tyvynain, Unifyer, Untrue Believer, Urhixidur, VanishedUser314159, Ventifact, Vicki Rosenzweig, VinnyXY, Viva-Verdi, Vladdy321, Vsmith, W4chris, Waffle411, Wayne Slam, Whisky drinker, WikHead, Wiki alf, Williamb, Willking1979, Wilson44691, Wimt, Wmahan, Woohookitty, Wtmitchell, Xiner, Yamara, Yansa, Zamphuor, فلى فى ع, 682 anonymous edits

Kimmeridge Clay *Source*: http://en.wikipedia.org/w/index.php?title=Kimmeridge_Clay *Contributors*: A Karley, Abyssal, Ballista, Chris the speller, Cj1340, CommonsDelinker, Gilesmorant, Grahbudd, Hovden, Invertzoo, J. Spencer, MWAK, Obsidian Soul, Paul H., Plazak, Rich Farmbrough, Snek01, Vsmith, Zundark, 5 anonymous edits

George Rolleston *Source*: http://en.wikipedia.org/w/index.php?title=George_Rolleston *Contributors*: Astrochemist, BD2412, Bencherlite, Bobblehead, DMS, DuncanHill, Giler, Grafen, Macdonald-ross, Materialscientist, Necrothesp, Occuli, Plucas58, Pruneau, Rjwilmsi, SE7, Timbouctou, Timrollpickering, 7 anonymous edits

Image Sources, Licenses and Contributors

file:Hadrosauroids.jpg *Source*: http://en.wikipedia.org/w/index.php?title=File:Hadrosauroids.jpg *License*: unknown *Contributors*: Pavel Riha = user Pavel.Riha.CB (e-mail)

File:Red Pencil Icon.png *Source*: http://en.wikipedia.org/w/index.php?title=File:Red_Pencil_Icon.png *License*: unknown *Contributors*: User:Peter coxhead

Image:Triceratops AMNH 01.jpg *Source*: http://en.wikipedia.org/w/index.php?title=File:Triceratops_AMNH_01.jpg *License*: unknown *Contributors*: Michael Gray from Wantagh NY, USA

File:Stegosaurus at the Field Museum.jpg *Source*: http://en.wikipedia.org/w/index.php?title=File:Stegosaurus_at_the_Field_Museum.jpg *License*: unknown *Contributors*: Original uploader was Killdevil at en.wikipedia

File:Edmontonia dinosaur.png *Source*: http://en.wikipedia.org/w/index.php?title=File:Edmontonia_dinosaur.png *License*: unknown *Contributors*: user:LadyofHats

File:Sprawling and erect hip joints - horiz.png *Source*: http://en.wikipedia.org/w/index.php?title=File:Sprawling_and_erect_hip_joints_-_horiz.png *License*: unknown *Contributors*: Original uploader was Philcha at en.wikipedia (Original text : Philcha (talk))

Image:Marasuchus.JPG *Source*: http://en.wikipedia.org/w/index.php?title=File:Marasuchus.JPG *License*: unknown *Contributors*: HoopoeBaijiKite Original uploader was Mitternacht90 at en.wikipedia

Image:Herrerasaurusskeleton.jpg *Source*: http://en.wikipedia.org/w/index.php?title=File:Herrerasaurusskeleton.jpg *License*: unknown *Contributors*: Zach Tirrell from Plymouth, USA

Image:Saurischia.png *Source*: http://en.wikipedia.org/w/index.php?title=File:Saurischia.png *License*: unknown *Contributors*: FunkMonk, Koobak, Muriel Gottrop

Image:Tyrannosaurus pelvis left.jpg *Source*: http://en.wikipedia.org/w/index.php?title=File:Tyrannosaurus_pelvis_left.jpg *License*: unknown *Contributors*: User:Ballista

Image:Ornithischia.png *Source*: http://en.wikipedia.org/w/index.php?title=File:Ornithischia.png *License*: unknown *Contributors*: Koobak, Muriel Gottrop, 2 anonymous edits

Image:Edmontosaurus pelvis left.jpg *Source*: http://en.wikipedia.org/w/index.php?title=File:Edmontosaurus_pelvis_left.jpg *License*: unknown *Contributors*: Conty, Kevmin, Smeira, Timichal

Image:Macronaria scrubbed enh.jpg *Source*: http://en.wikipedia.org/w/index.php?title=File:Macronaria_scrubbed_enh.jpg *License*: unknown *Contributors*: Original uploader was Killdevil at en.wikipedia

Image:Ornithopods jconway.jpg *Source*: http://en.wikipedia.org/w/index.php?title=File:Ornithopods_jconway.jpg *License*: unknown *Contributors*: Dudo, FunkMonk, G.dallorto, John.Conway, Kevmin, Koobak

Image:Largestdinosaursbysuborder scale.png *Source*: http://en.wikipedia.org/w/index.php?title=File:Largestdinosaursbysuborder_scale.png *License*: unknown *Contributors*: User:Dinoguy2

Image:Giraffatitan scale.png *Source*: http://en.wikipedia.org/w/index.php?title=File:Giraffatitan_scale.png *License*: unknown *Contributors*: User:Dinoguy2

Image:Human-eoraptor size comparison(v2).png *Source*: http://en.wikipedia.org/w/index.php?title=File:Human-eoraptor_size_comparison(v2).png *License*: unknown *Contributors*: User:Dropzink

Image:Maiasaurusnest.jpg *Source*: http://en.wikipedia.org/w/index.php?title=File:Maiasaurusnest.jpg *License*: unknown *Contributors*: Dudo, Kevmin, Koobak

Image:Dino eggP9240092.JPG *Source*: http://en.wikipedia.org/w/index.php?title=File:Dino_eggP9240092.JPG *License*: unknown *Contributors*: user:deror_avi

Image:Centrosaurus dinosaur.png *Source*: http://en.wikipedia.org/w/index.php?title=File:Centrosaurus_dinosaur.png *License*: unknown *Contributors*: user:LadyofHats

Image:Palais de la Decouverte Tyrannosaurus rex p1050042.jpg *Source*: http://en.wikipedia.org/w/index.php?title=File:Palais_de_la_Decouverte_Tyrannosaurus_rex_p1050042.jpg *License*: unknown *Contributors*: User:David.Monniaux

Image:Eubrontes01.JPG *Source*: http://en.wikipedia.org/w/index.php?title=File:Eubrontes01.JPG *License*: unknown *Contributors*: User:Wilson44691

Image:SArchaeopteryxBerlin2.jpg *Source*: http://en.wikipedia.org/w/index.php?title=File:SArchaeopteryxBerlin2.jpg *License*: unknown *Contributors*: Owen, Richard

File:Pneumatopores on the left ilium of the theropod Aerosteon riocoloradensis.jpg *Source*: http://en.wikipedia.org/w/index.php?title=File:Pneumatopores_on_the_left_ilium_of_the_theropod_Aerosteon_riocoloradensis.jpg *License*: unknown *Contributors*: Sereno PC, Martinez RN, Wilson JA, Varricchio DJ, Alcober OA, et al.

Image:Chicxulub radar topography.jpg *Source*: http://en.wikipedia.org/w/index.php?title=File:Chicxulub_radar_topography.jpg *License*: unknown *Contributors*: NASA/JPL-Caltech

Image:Stego-marsh-1896-US geological survey.png *Source*: http://en.wikipedia.org/w/index.php?title=File:Stego-marsh-1896-US_geological_survey.png *License*: unknown *Contributors*: User:Anetode

Image:William Buckland detail.png *Source*: http://en.wikipedia.org/w/index.php?title=File:William_Buckland_detail.png *License*: unknown *Contributors*: Original uploader was Killdevil at en.wikipedia

Image:OthnielCharlesMarsh.jpg *Source*: http://en.wikipedia.org/w/index.php?title=File:OthnielCharlesMarsh.jpg *License*: unknown *Contributors*: Kilom691, Librotyrannus, Shizhao, Väsk, 1 anonymous edits

Image:edcope.jpg *Source*: http://en.wikipedia.org/w/index.php?title=File:Edcope.jpg *License*: unknown *Contributors*: F. Gutekunst UNIQ-ref-1-0f28921bfd9e0227-QINU

File:Biological_classification_L_Pengo_vflip.svg *Source*: http://en.wikipedia.org/w/index.php?title=File:Biological_classification_L_Pengo_vflip.svg *License*: unknown *Contributors*: Adrignola, ArnoLagrange, Nisetpdajsankha, Pavel55, Pengo

File:Geologic Clock with events and periods.svg *Source*: http://en.wikipedia.org/w/index.php?title=File:Geologic_Clock_with_events_and_periods.svg *License*: unknown *Contributors*: User:Hardwigg, User:Woudloper

Image:TectonicReconstructionGlobal2.gif *Source*: http://en.wikipedia.org/w/index.php?title=File:TectonicReconstructionGlobal2.gif *License*: unknown *Contributors*: User:Aineias

Image:Geological time spiral.png *Source*: http://en.wikipedia.org/w/index.php?title=File:Geological_time_spiral.png *License*: unknown *Contributors*: United States Geological Survey

Image:Geological Time Scale.png *Source*: http://en.wikipedia.org/w/index.php?title=File:Geological_Time_Scale.png *License*: unknown *Contributors*: Richard S. Murphy, Jr.

Image:Dacentrurus.png *Source*: http://en.wikipedia.org/w/index.php?title=File:Dacentrurus.png *License*: unknown *Contributors*: Original uploader was Smokeybjb at en.wikipedia

File:Trigonellites latus.jpg *Source*: http://en.wikipedia.org/w/index.php?title=File:Trigonellites_latus.jpg *License*: unknown *Contributors*: Cooke, A. H., Shipley, A. E. & Reed, F. R. C.

File:George Rolleston2.jpg *Source*: http://en.wikipedia.org/w/index.php?title=File:George_Rolleston2.jpg *License*: unknown *Contributors*: Wheeler and Day (London)

File:George Rolleston.jpg *Source*: http://en.wikipedia.org/w/index.php?title=File:George_Rolleston.jpg *License*: unknown *Contributors*: User:Pruneau

GNU Free Documentation License Version 1.2, November 2002 Copyright (C) 2000,2001,2002 Free Software Foundation, Inc. 59 Temple Place, Suite 330, Boston, MA 02111-1307 USA Everyone is permitted to copy and distribute verbatim copies of this license document, but changing it is not allowed.

0. PREAMBLE

The purpose of this License is to make a manual, textbook, or other functional and useful document "free" in the sense of freedom: to assure everyone the effective freedom to copy and redistribute it, with or without modifying it, either commercially or noncommercially. Secondarily, this License preserves for the author and publisher a way to get credit for their work, while not being considered responsible for modifications made by others. This License is a kind of "copyleft", which means that derivative works of the document must themselves be free in the same sense. It complements the GNU General Public License, which is a copyleft license designed for free software. We have designed this License in order to use it for manuals for free software, because free software needs free documentation: a free program should come with manuals providing the same freedoms that the software does. But this License is not limited to software manuals; it can be used for any textual work, regardless of subject matter or whether it is published as a printed book. We recommend this License principally for works whose purpose is instruction or reference.

1. APPLICABILITY AND DEFINITIONS

This License applies to any manual or other work, in any medium, that contains a notice placed by the copyright holder saying it can be distributed under the terms of this License. Such a notice grants a world-wide, royalty-free license, unlimited in duration, to use that work under the conditions stated herein. The "Document", below, refers to any such manual or work. Any member of the public is a licensee, and is addressed as "you". You accept the license if you copy, modify or distribute the work in a way requiring permission under copyright law. A "Modified Version" of the Document means any work containing the Document or a portion of it, either copied verbatim, or with modifications and/or translated into another language. A "Secondary Section" is a named appendix or a front-matter section of the Document that deals exclusively with the relationship of the publishers or authors of the Document to the Document's overall subject (or to related matters) and contains nothing that could fall directly within that overall subject. (Thus, if the Document is in part a textbook of mathematics, a Secondary Section may not explain any mathematics.) The relationship could be a matter of historical connection with the subject or with related matters, or of legal, commercial, philosophical, ethical or political position regarding them. The "Invariant Sections" are certain Secondary Sections whose titles are designated, as being those of Invariant Sections, in the notice that says that the Document is released under this License. If a section does not fit the above definition of Secondary then it is not allowed to be designated as Invariant. The Document may contain zero Invariant Sections. If the Document does not identify any Invariant Sections then there are none. The "Cover Texts" are certain short passages of text that are listed, as Front-Cover Texts or Back-Cover Texts, in the notice that says that the Document is released under this License. A Front-Cover Text may be at most 5 words, and a Back-Cover Text may be at most 25 words. A "Transparent" copy of the Document means a machine-readable copy, represented in a format whose specification is available to the general public, that is suitable for revising the document straightforwardly with generic text editors or (for images composed of pixels) generic paint programs or (for drawings) some widely available drawing editor, and that is suitable for input to text formatters or for automatic translation to a variety of formats suitable for input to text formatters. A copy made in an otherwise Transparent file format whose markup, or absence of markup, has been arranged to thwart or discourage subsequent modification by readers is not Transparent. An image format is not Transparent if used for any substantial amount of text. A copy that is not "Transparent" is called "Opaque". Examples of suitable formats for Transparent copies include plain ASCII without markup, Texinfo input format, LaTeX input format, SGML or XML using a publicly available DTD, and standard-conforming simple HTML, PostScript or PDF designed for human modification. Examples of transparent image formats include PNG, XCF and JPG. Opaque formats include proprietary formats that can be read and edited only by proprietary word processors, SGML or XML for which the DTD and/or processing tools are not generally available, and the machine-generated HTML, PostScript or PDF produced by some word processors for output purposes only. The "Title Page" means, for a printed book, the title page itself, plus such following pages as are needed to hold, legibly, the material this License requires to appear in the title page. For works in formats which do not have any title page as such, "Title Page" means the text near the most prominent appearance of the work's title, preceding the beginning of the body of the text. A section "Entitled XYZ" means a named subunit of the Document whose title either is precisely XYZ or contains XYZ in parentheses following text that translates XYZ in another language. (Here XYZ stands for a specific section name mentioned below, such as "Acknowledgements", "Dedications", "Endorsements", or "History".) To "Preserve the Title" of such a section when you modify the Document means that it remains a section "Entitled XYZ" according to this definition. The Document may include Warranty Disclaimers next to the notice which states that this License applies to the Document. These Warranty Disclaimers are considered to be included by reference in this License, but only as regards disclaiming warranties: any other implication that these Warranty Disclaimers may have is void and has no effect on the meaning of this License.

2. VERBATIM COPYING

You may copy and distribute the Document in any medium, either commercially or noncommercially, provided that this License, the copyright notices, and the license notice saying this License applies to the Document are reproduced in all copies, and that you add no other conditions whatsoever to those of this License. You may not use technical measures to obstruct or control the reading or further copying of the copies you make or distribute. However, you may accept compensation in exchange for copies. If you distribute a large enough number of copies you must also follow the conditions in section 3. You may also lend copies, under the same conditions stated above, and you may publicly display copies.

3. COPYING IN QUANTITY

If you publish printed copies (or copies in media that commonly have printed covers) of the Document, numbering more than 100, and the Document's license notice requires Cover Texts, you must enclose the copies in covers that carry, clearly and legibly, all these Cover Texts: Front-Cover Texts on the front cover, and Back-Cover Texts on the back cover. Both covers must also clearly and legibly identify you as the publisher of these copies. The front cover must present the full title with all words of the title equally prominent and visible. You may add other material on the covers in addition. Copying with changes limited to the covers, as long as they preserve the title of the Document and satisfy these conditions, can be treated as verbatim copying in other respects. If the required texts for either cover are too voluminous to fit legibly, you should put the first ones listed (as many as fit reasonably) on the actual cover, and continue the rest onto adjacent pages. If you publish or distribute Opaque copies of the Document numbering more than 100, you must either include a machine-readable Transparent copy along with each Opaque copy, or state in or with each Opaque copy a computer-network location from which the general network-using public has access to download using public-standard network protocols a complete Transparent copy of the Document, free of added material. If you use the latter option, you must take reasonably prudent steps, when you begin distribution of Opaque copies in quantity, to ensure that this Transparent copy will remain thus accessible at the stated location until at least one year after the last time you distribute an Opaque copy (directly or through your agents or retailers) of that edition to the public. It is requested, but not required, that you contact the authors of the Document well before redistributing any large number of copies, to give them a chance to provide you with an updated version of the Document.

4. MODIFICATIONS

You may copy and distribute a Modified Version of the Document under the conditions of sections 2 and 3 above, provided that you release the Modified Version under precisely this License, with the Modified Version filling the role of the Document, thus licensing distribution and modification of the Modified Version to whoever possesses a copy of it. In addition, you must do these things in the Modified Version: A. Use in the Title Page (and on the covers, if any) a title distinct from that of the Document, and from those of previous versions (which should, if there were any, be listed in the History section of the Document). You may use the same title as a previous version if the original publisher of that version gives permission. B. List on the Title Page, as authors, one or more persons or entities responsible for authorship of the modifications in the Modified Version, together with at least five of the principal authors of the Document (all of its principal authors, if it has fewer than five), unless they release you from this requirement. C. State on the Title page the name of the publisher of the Modified Version, as the publisher. D. Preserve all the copyright notices of the Document. E. Add an appropriate copyright notice for your modifications adjacent to the other copyright notices. F. Include, immediately after the copyright notices, a license notice giving the public permission to use the Modified Version under the terms of this License, in the form shown in the Addendum below. G. Preserve in that license notice the full lists of Invariant Sections and required Cover Texts given in the Document's license notice. H. Include an unaltered copy of this License. I. Preserve the section Entitled "History", Preserve its Title, and add to it an item stating at least the title, year, new authors, and publisher of the Modified Version as given on the Title Page. If there is no section Entitled "History" in the Document, create one stating the title, year, authors, and publisher of the Document as given on its Title Page, then add an item describing the Modified Version as stated in the previous sentence. J. Preserve the network location, if any, given in the Document for public access to a Transparent copy of the Document, and likewise the network locations given in the Document for previous versions it was based on. These may be placed in the "History" section. You may omit a network location for a work that was published at least four years before the Document itself, or if the original publisher of the version it refers to gives permission. K. For any section Entitled "Acknowledgements" or "Dedications", Preserve the Title of the section, and preserve in the section all the substance and tone of each of the contributor acknowledgements and/or dedications given therein. L. Preserve all the Invariant Sections of the Document, unaltered in their text and in their titles. Section numbers or the equivalent are not considered part of the section titles. M. Delete any section Entitled "Endorsements". Such a section may not be included in the Modified Version. N. Do not retitle any existing section to be Entitled "Endorsements" or to conflict in title with any Invariant Section. O. Preserve any Warranty Disclaimers. If the Modified Version includes new front-matter sections or appendices that qualify as Secondary Sections and contain no material copied from the Document, you may at your option designate some or all of these sections as invariant. To do this, add their titles to the list of Invariant Sections in the Modified Version's license notice. These titles must be distinct from any other section titles. You may add a section Entitled "Endorsements", provided it contains nothing but endorsements of your Modified Version by various parties--for example, statements of peer review or that the text has been approved by an organization as the authoritative definition of a standard. You may add a passage of up to five words as a Front-Cover Text, and a passage of up to 25 words as a Back-Cover Text, to the end of the list of Cover Texts in the Modified Version. Only one passage of Front-Cover Text and one of Back-Cover Text may be added by (or through arrangements made by) any one entity. If the Document already includes a cover text for the same cover, previously added by you or by arrangement made by the same entity you are acting on behalf of, you may not add another; but you may replace the old one, on explicit permission from the previous publisher that added the old one. The author(s) and publisher(s) of the Document do not by this License give permission to use their names for publicity for or to assert or imply endorsement of any Modified Version.

5. COMBINING DOCUMENTS

You may combine the Document with other documents released under this License, under the terms defined in section 4 above for modified versions, provided that you include in the combination all of the Invariant Sections of all of the original documents, unmodified, and list them all as Invariant Sections of your combined work in its license notice, and that you preserve all their Warranty Disclaimers. The combined work need only contain one copy of this License, and multiple identical Invariant Sections may be replaced with a single copy. If there are multiple Invariant Sections with the same name but different contents, make the title of each such section unique by adding at the end of it, in parentheses, the name of the original author or publisher of that section if known, or else a unique number. Make the same adjustment to the section titles in the list of Invariant Sections in the license notice of the combined work. In the combination, you must combine any sections Entitled "History" in the various original documents, forming one section Entitled "History"; likewise combine any sections Entitled "Acknowledgements", and any sections Entitled "Dedications". You must delete all sections Entitled "Endorsements".

6. COLLECTIONS OF DOCUMENTS

You may make a collection consisting of the Document and other documents released under this License, and replace the individual copies of this License in the various documents with a single copy that is included in the collection, provided that you follow the rules of this License for verbatim copying of each of the documents in all other respects. You may extract a single document from such a collection, and distribute it individually under this License, provided you insert a copy of this License into the extracted document, and follow this License in all other respects regarding verbatim copying of that document.

7. AGGREGATION WITH INDEPENDENT WORKS

A compilation of the Document or its derivatives with other separate and independent documents or works, in or on a volume of a storage or distribution medium, is called an "aggregate" if the copyright resulting from the compilation is not used to limit the legal rights of the compilation's users beyond what the individual works permit. When the Document is included in an aggregate, this License does not apply to the other works in the aggregate which are not themselves derivative works of the Document. If the Cover Text requirement of section 3 is applicable to these copies of the Document, then if the Document is less than one half of the entire aggregate, the Document's Cover Texts may be placed on covers that bracket the Document within the aggregate, or the electronic equivalent of covers if the Document is in electronic form. Otherwise they must appear on printed covers that bracket the whole aggregate.

8. TRANSLATION

Translation is considered a kind of modification, so you may distribute translations of the Document under the terms of section 4. Replacing Invariant Sections with translations requires special permission from their copyright holders, but you may include translations of some or all Invariant Sections in addition to the original versions of these Invariant Sections. You may include a translation of this License, and all the license notices in the Document, and any Warranty Disclaimers, provided that you also include the original English version of this License and the original versions of those notices and disclaimers. In case of a disagreement between the translation and the original version of this License or a notice or disclaimer, the original version will prevail. If a section in the Document is Entitled "Acknowledgements", "Dedications", or "History", the requirement (section 4) to Preserve its Title (section 1) will typically require changing the actual title.

9. TERMINATION

You may not copy, modify, sublicense, or distribute the Document except as expressly provided for under this License. Any other attempt to copy, modify, sublicense or distribute the Document is void, and will automatically terminate your rights under this License. However, parties who have received copies, or rights, from you under this License will not have their licenses terminated so long as such parties remain in full compliance.

10. FUTURE REVISIONS OF THIS LICENSE

The Free Software Foundation may publish new, revised versions of the GNU Free Documentation License from time to time. Such new versions will be similar in spirit to the present version, but may differ in detail to address new problems or concerns. See http://www.gnu.org/copyleft/. Each version of the License is given a distinguishing version number. If the Document specifies that a particular numbered version of this License "or any later version" applies to it, you have the option of following the terms and conditions either of that specified version or of any later version that has been published (not as a draft) by the Free Software Foundation. If the Document does not specify a version number of this License, you may choose any version ever published (not as a draft) by the Free Software Foundation. ADDENDUM: How to use this License for your documents To use this License in a document you have written, include a copy of the License in the document and put the following copyright and license notices just after the title page: Copyright (c) YEAR YOUR NAME. Permission is granted to copy, distribute and/or modify this document under the terms of the GNU Free Documentation License, Version 1.2 or any later version published by the Free Software Foundation; with no Invariant Sections, no Front-Cover Texts, and no Back-Cover Texts. A copy of the license is included in the section entitled "GNU Free Documentation License". If you have Invariant Sections, Front-Cover Texts and Back-Cover Texts, replace the "with...Texts." line with this: with the Invariant Sections being LIST THEIR TITLES, with the Front-Cover Texts being LIST, and with the Back-Cover Texts being LIST. If you have Invariant Sections without Cover Texts, or some other combination of the three, merge those two alternatives to suit the situation. If your document contains nontrivial examples of program code, we recommend releasing these examples in parallel under your choice of free software license, such as the GNU General Public License, to permit their use in free software.

Printed by Books on Demand GmbH, Norderstedt / Germany